Textile Progress

2008   Vol 40   No 3

Advanced technical textile products

T. Matsuo

The Textile Institute

SUBSCRIPTION INFORMATION

Textile Progress (USPS Permit Number pending), Print ISSN 0040-5167, Online ISSN 1754-2278, Volume 40, 2008.

Textile Progress (www.informaworld.com/textileprogress) is a peer-reviewed journal published quarterly in March, June, September and December by Taylor & Francis, 4 Park Square, Milton Park, Abingdon, Oxon, OX14 4RN, UK on behalf of The Textile Institute.

Institutional Subscription Rate (print and online): $304/£160/€243
Institutional Subscription Rate (online-only): $288/£152/€230 (plus tax where applicable)
Personal Subscription Rate (print only): $112/£58/€90

All current institutional subscriptions include online access for any number of concurrent users across a local area network to the currently available backfile and articles posted online ahead of publication.

Ordering Information: Please contact your local Customer Service Department to take out a subscription to the Journal: **India**: Universal Subscription Agency Pvt. Ltd, 101–102 Community Centre, Malviya Nagar Extn, Post Bag No. 8, Saket, New Delhi 110017. **Japan**: Kinokuniya Company Ltd, Journal Department, PO Box 55, Chitose, Tokyo 156. **USA, Canada and Mexico**: Taylor & Francis, 325 Chestnut Street, 8th Floor, Philadelphia, PA 19106, USA. Tel: +1 800 354 1420 or +1 215 625 8900; fax: +1 215 625 8914, email: customerservice@taylorandfrancis.com. **UK and all other territories**: T&F Customer Services, Informa Plc., Sheepen Place, Colchester, Essex, CO3 3LP, UK. Tel: +44 (0)20 7017 5544; fax: +44 (0)20 7017 5198, email: tf.enquiries@tfinforma.com.

Dollar rates apply to all subscribers outside Europe. Euro rates apply to all subscribers in Europe, except the UK and the Republic of Ireland where the pound sterling price applies. All subscriptions are payable in advance and all rates include postage. Journals are sent by air to the USA, Canada, Mexico, India, Japan and Australasia. Subscriptions are entered on an annual basis, i.e. January to December. Payment may be made by sterling cheque, dollar cheque, euro cheque, international money order, National Giro or credit cards (Amex, Visa and Mastercard).

Back Issues: Taylor & Francis retains a three year back issue stock of journals. Older volumes are held by our official stockists to whom all orders and enquiries should be addressed:
Periodicals Service Company, 11 Main Street, Germantown, NY 12526, USA. Tel: +1 518 537 4700; fax: +1 518 537 5899; email: psc@periodicals.com.

Periodical postage paid at Jamaica, NY 11431, by US Mailing Agent Air Business Ltd, c/o Worldnet Shipping USA Inc., 149-35 177th Street, Jamaica, New York, NY 11434.

Subscription records are maintained at Taylor & Francis Group, 4 Park Square, Milton Park, Abingdon, OX14 4RN, United Kingdom.

For more information on Taylor & Francis' journal publishing program, please visit our website: www.informaworld.com/journals.

CONTENTS

Textile Progress
Vol. 40, No. 3, 2008, 123–181

Taylor & Francis
Taylor & Francis Group

Advanced technical textile products

Tatsuki Matsuo*

SCI-TEX, 12-15 Hanazono-cho, Ohtsu, 520-022, Japan

(Received 5 August 2008; final version received 1 September 2008)

This article is situated to be successive to "Fibre materials for advanced technical textiles" in the series of "Advanced technical textiles" of Textile Progress. In the previous article, fiber materials used for advanced technical textiles are introduced. In this article, advanced technical textiles products are described according to the application fields of the fiber materials. Although this article does not cover all the end-uses, it contains major parts of advanced technical textile products, which include products for resources and environmental issues, for automobiles, for medical uses, for protective uses, for information technologies, for civil engineering and for electronics textiles.

Keywords: advanced technical textile products, resources and environmental issue, automobiles, medical uses, protection uses, information technologies, civil engineering, electronic textiles

1. Introduction

1.1 The content and objectives of this book in terms of advanced technical textiles series

In the former article in this advanced technical textile series, fiber materials for advanced technical textiles [1] have been published. In this article, advanced technical textile products, which are the application products of these fiber materials are introduced from the technological point of view. The application fields in this article are classified as (1) resources and environmental matters, (2) automobile, (3) medical and hygienic, (4) protection and safety, (5) electric and information technologies, (6) construction and civil engineering, and (7) E-textiles.

There are some application fields of advanced technical textiles to be added to the above (1) to (7) items, such as sports, agriculture/sericulture, marine/fishery, packaging, and textiles for production machines and systems. But they are not included in this article to save the pages under the volume restriction. Although electronic textiles are still at an infancy stage, they have now become one of the exciting application fields as to be treated in an independent chapter.

Objectives in the concept for writing this article are as follows:

(a) to provide a wide view and systematic/comprehensive source of knowledge on advanced technical textile products to young professionals and graduate students,
(b) to give them an understanding of how important advanced technical textile products are meant for our welfare and for the other fields of industries as key materials, and

---

*Email: tamatsuo@nifty.com

ISSN 0040-5167 print/ISSN 1754-2278 online
DOI: 10.1080/00405160802386063
http://www.informaworld.com

(c) to give them a feeling of how large the potential of advanced technical textile products is in its technological advancement.

The topics in the article must cover quite a wide range to cope with the above objectives within the volume restriction. Therefore, all the description must be compact. Hence, the readers may often want to know more in detail. In addition, many cited references in this article are written only in Japanese. In order to avoid this difficulty to some extent, a few important books related to technical textiles written in English are especially listed at the end of the "references" part.

1.2 Advanced technical textiles and their structural features

In this article, the term *technical textiles* is used as "textiles for nonapparel, nonhousehold/furnishing end uses, whose values are highly based on their technical performance and functional properties." The term *advanced technical textiles* in this article means "technical textiles that have some technological advancement in the material and/or the application."

Generally, the reason why textile products are selectively used for a certain specified end use instead of other kinds of materials is that the textile products in which some of the configurational functions of fiber are effectively utilized can become optimal in the ratio (performance/cost), among several forms of materials. The configurational functions of fiber are consisted of the following four elements:

(1) It is flexible (pliable).
(2) It has high ability in its axial transmission of such properties as mechanical load.
(3) It has high specific surface area.
(4) It has technological easiness in transformability into textile structural materials such as weaves and nonwovens.

In many cases of fiber-reinforced composites, functional elements (2), (3), and (4) are fully utilized. In several kinds of membrane hollow fibers, the element (3) is highly utilized. Optical fiber utilizes firstly the element (2) and secondly the element (1). In tyres, (1), (2), and (3) are main functional elements.

The required performance for textile products is varied with detailed kinds of end uses. Then their optimum structures are fully dependent on their end uses. Therefore, technical textile products are much wider in their structural range than apparel textiles.

1.3 Industrial considerations of advanced technical textiles

1.3.1 The features of business in advanced technical textiles

Most advanced technical textiles are commercialized by some technological development to fill a specific requirement in the product, from a specific customer. The development is usually conducted by a manufacturer in cooperation with the customer whose role is to evaluate trial sample(s) of applied product from the viewpoint of the ratio (performance/cost). In some cases, the cooperation by its raw material supplier and/or its final manufacturer is needed. In these cases, the manufacturer can be generally a kind of pioneer in that technological point. After the success in the commercialization of the developed product, the base of the technology thus obtained can be utilized for other similar requirements of that customer and/or the other customers. Therefore, the next activity to increase the sale volume of textiles related to the developed technology is to effectively find other such new customer requirements. By repeating this kind of evolving activity, the sale of the products can be gradually increased. As understood by these descriptions, the heart of some strong enthu-siasm and patience are generally needed for the manufacturer in the business of advanced technical textiles. In contrast to conventional textile business, the life of commercial value for the advanced technical textile product thus developed is usually much longer because of the technological barrier grown by the development.

In the business of advanced technical textiles, the manufacturer must be sensitive to patent management. In the early stages of the development, he must carefully check if all the technologies in the development will not be obstructed by any existing patents and must also apply the patents related to the development to guard the development and the output products.

1.3.2 The market size of advanced technical textiles

The product forms of technical textiles for a specified manufacturer can be fiber, textiles (fabrics etc.) or unit/parts using textile. Statistics of amount for technical textiles are influenced by what product form is adopted for the statistics and how technical textiles are defined in their range for the statistics. Hence market sizes of technical textiles in amount and value are not so clear. But Table 1, which shows worldwide consumption statistics of technical textiles reported from an investigation company [2], must be useful to have a rough idea of the market size of technical textiles by application field. There are no statistics for advanced technical textiles, which is fully dependent on the range defined as advanced technical textiles. But roughly speaking, the author feels that their share in technical textiles is over 50% in advanced countries in total size.

Table 1.   Worldwide consumption of technical textiles by application fields [2].

	10^3 Tonnes			$ Million		
	2000	2005	Growth (% per annum)	2000	2005	Growth (% per annum)
Transport textiles (auto, train, sea, aero)	2,220	2,480	2.2	13,080	14,370	1.9
Industrial products and components	1,880	2,340	4,5	9,290	11,560	4.5
Medical and hygiene textiles	1,380	1,650	3.6	7,820	9,530	4.0
Home textiles, domestic equipment	1,800	2,260	4.7	7,780	9,680	4.5
Clothing components (thread, interlinings)	730	820	2.3	6,800	7,640	2.4
Agriculture, horticulture and fishing	900	1,020	2.5	4,260	4,940	3.0
Construction – building and roofing	1,030	1,270	4.3	3,390	4,320	5.0
Packaging and containment	530	660	4.5	2,320	2,920	4.7
Sport and leisure (excluding apparel)	310	390	4.7	2,030	2,510	4.3
Geotextiles, civil engineering	400	570	7.3	1,860	2,660	7.4
Protective and safety clothing and textiles	160	220	6.6	1,640	2,230	6.3
Total above	11,340	13,680	3.9	60,270	72,360	3.7
Ecological protection textiles	230	310	6.2	1,270	1,610	4.9

1.3.3 The industrial situation of advanced technical textiles

The amount of total fiber consumption in the world is gradually growing with a rate little higher than the increase of population. As shown in Table 1, the growth rate of technical textiles in the world is 3.7%, which is a little higher than that of total fiber consumption. In worldwide fiber consumption amount, the share of technical textiles is estimated to be about 20%. That figure is about 40% in the advanced countries. But it is also noted that the growth rate of technical textiles in developing countries, especially China, is much higher than that in advanced countries. It means that the importance of technical textiles in the total textile industry is also increasing in developing countries.

Some applications are hygienic and medical in which technical textiles can directly contribute to human welfare. But most technical textile products are supplied as materials or the units/parts to the other kinds of industries such as automobile industry, as naturally understood by the above description. Particulary in the case of advanced technical textile products, they are often one of the key materials or parts/units for realizing the performance of end-use products.

1.3.4 The trends and future scope of advanced technical textiles

Historically, clothing has always been the major application area of textiles in the early stages of textile industry. One of the important factors behind the growth of technical textiles is the appearance of man-made fibers. Another factor is an increase in the industrial or social needs for technical textiles. The specific additionally necessary condition for the industry of manufacturing advanced technical textiles is an industrial base for developing advanced technical textiles. Hence the industry of advanced technical textiles has grown in advanced countries. Now it is gradually spreading to developing countries. In the textile industry of advanced countries, the trend orienting to advanced technical textiles has become significant.

Now the problems of global environment and resources have been more serious. The needs for convenience, comfort, health, and safety have increased. These trends will surely present several new opportunities for the business of advanced technical textiles. Therefore, advanced technical textiles will gradually and steadily increase in the future.

2. Textile products contributing to resources and environmental issues

2.1 General scope of textile products contributing to resources and environmental issues

Problems of global environment and resources have become one of the most serious issues for man. The rise in oil prices directly causes the cost of making synthetic fibers to move up. Even natural and man-made fibers that originate from plants are significantly influenced by oil prices. On the other hand, there is a big space for fiber and textile technologies to contribute toward this issue. It means that there can be a big business opportunity for the fiber/textile industry.

It is believed that the technologies can be conveniently classified into defensive technologies and offensive technologies for fiber/textile industry. The former are those that can contribute to reduce the loads caused by the textile industry on global environment and resources. The latter are those that can contribute to reduce the loads originating from the other kinds of industries and human living/activities. Such technologies as (1) biodegradable fibers, (2) fibers/textiles made with low energy consumption and no environmentally harmful exhaust substances, (3) recycling systems for textile wastes, (4) waste reduction and energy saving at fiber plant/textile mills, (5) treatment of wastes at fiber plant/textile mills, and (6) reuse and recycling of textile products can be classified as defensive. Offensive technologies can be mentioned as follows: (1) water treatment and purification; (2) air purification; (3) enhancement of energy resources through fibrous materials for the electrode and the separator of battery/fuel cell, for reinforcement of wind turbine and through composites for lightening transport vehicles; (4) energy saving through comfortable wears for mild cool air conditioning in summer and for mild warm air conditioning in winter; (5) energy saving through fibrous materials by thermal insulation; (6) recycling of polyethylene terephthalate (PET) bottle into fibrous materials; and (7) water-saving systems for plantation.

Regarding the items (1) and (2), some introductive explanations are given in the former article [1] in this series. The items (1), (2), (3), (4), and (5) in offensive technologies and item (6) as technical textiles in defensive technologies are introduced in this chapter.

2.2 Water treatment and purification

2.2.1 Water treatment by filtration and bioreaction using fibrous materials

2.2.1.1. Filtration systems. Sand filtration is widely used for water treatment. But some filtration systems using fibrous materials have been developed. The main purpose of such systems is to significantly reduce the area of filtration process. In a fibrous filtration system, filtration medium is a layer piled with a number of fiber-mass blocks. Periodically filtrated particles on the medium are removed by counterwashing accompanied by aeration. Figure 1 schematically shows one of noticeable systems [3] in which filtration is continuously carried out on the inner surface of a rotating drum. The filter medium is composed of a double-layered satin fabric whose surface is covered by raised microfibers of the weft yarn. Filtrated particles on the surface are removed into condensed wastewater by counterwashing. It is especially useful for the treatment of water containing large amount of planktons.

The uses of ultrafiltration (UF) and microfiltration (MF) membrane hollow fibers (see Figure 6 in the former article [1]) for nonbionic wastewater treatment have been increased. Large numbers of hollow fibers are contained as a bundle within a module. There are also other forms of UF/MF membrane such as tubular, spiral, and plane. But hollow fiber form is advantageous in unit compactness and simple handling over the other forms.

2.2.1.2. Bioreaction systems. In case that the filter medium is a layer piled by large number of fiber-mass blocks sustaining microorganism, the filter systems described in section

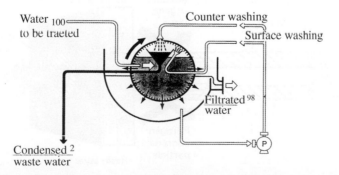

Figure 1. Filtration system using rotating drum covered by a microfiber woven fabric [3].

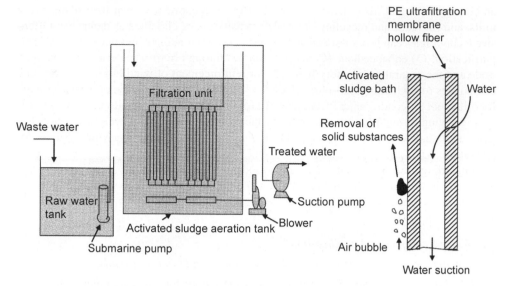

Figure 2. Bioreactor using membrane hollow fibers. (a) reactor [4] (b) actions to a hollow fiber; MBR indicates membrane bioreactor.

2.2.2.1 can act as not only a filtration vessel, but also as a bioreaction vessel to reduce the biological oxygen demand of organic wastewater. Figure 2 illustrates a bioreactor using UF-/MF-membrane hollow fiber. In the system, large numbers of the fiber module are dipped in activated sludge aeration tank. Water suction is applied in the hollow part of the fiber. Activated sludge filtrated on the outer surface of the fiber is removed by the bubble made by aeration. As compared to conventional bioreactor tank, it can be much more compact and the quality of treated water is much higher [4].

2.2.2 Reverse osmosis and UF/MF using membrane systems for the production of purified water

2.2.2.1. Reverse osmosis. Figure 3 shows the principle of reverse osmosis (RO) hollow fiber, whose membrane is nonporous (see Figure 6 in the former article [1]). By applying higher pressure than the RO pressure, water permeates through the membrane into its hollow part. Such materials as salt, virus, and pyrogen can be rejected at the surface of

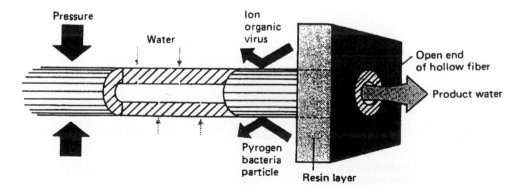

Figure 3. Principle of reverse osmosis by membrane hollow fiber.

Figure 4. An example of RO desalination plant from seawater [5].

active separation layer in membrane. Figure 4 shows an example of an RO desalination plant [5]. Large numbers of RO modules are installed, which contain hollow-fiber bundle. Another type of RO membrane than the hollow fiber is planer membrane wound in spiral form, in which nonwoven is used for the substrate of membrane. The main end use of RO membrane is the production of drinkable water by the desalination of seawater.

2.2.2.2. UF and MF for water work. In water work, the requirement to remove pathogenic microorganisms has been enhanced, especially in developed countries. Most of UF- and MF-membrane hollow fibers are very feasible for the requirement (see Figure 6 in the former article [1]).

2.2.2.3. Reuse and recycling of water. In order to efficiently utilize water, several technological systems for the reuse of wastewater or water recycling have intensively developed. In these systems, RO-, UF- and MF-membrane hollow fibers have been utilized according to water-quality requirements

2.2.3 Water purification by some specific fibers

2.2.3.1. Activated carbon fiber. Activated carbon fiber is often used in combination with the UF-/MF-membrane hollow fiber as a major component of water-purifying equipment for drinking water. The main role of activated fiber in the equipment is to remove such substance as chlorinated organic chemical and smell constituents.

2.2.3.2. Photocatalyst fiber. A water purification system using conical nonwoven units made of photocatalyst fiber has been developed [6]. The constituent of the fiber is gradually varied from silica at the core part to TiO_2(photocatalyst) at the surface. In the system, an ultraviolet (UV) lamp is located at the axial center of the conical units line. Contaminants in the water are oxidized and degraded by the photocatalyst under the UV light during the passage of water through the conical units line. The system is going to be widely applied for purifying several kinds of water in recycled water usages such as public bath, spring bath, swimming pool and industrial use water.

2.2.3.3. Ion-exchangeable fibers. There are some kinds of ion-exchangeable fibers. They can be effective to remove toxic ingredients of heavy metal from water. The fiber made of ion-exchangeable polystyrene resin is used for purification of recycled water from atomic power plants.

2.2.4 Separation of oil and water

2.2.4.1. Separation and removal of oil from seawater. The flow out of oil into sea or river has often caused a serious environmental problem. Oil fence is used to limit the area of oil contamination on the water surface. It is typically composed of (a) sheet fence made of woven fabric reinforced by belt, (b) float made of foamed polystyrene covered with woven fabric, and (c) weight for stabilization.

Oil floating on water surface can be removed by making use of oil-adsorptive fibrous sheet. The material of this sheet is usually nonwoven made of polypropylene (PP), kapok, or cotton. It can take several forms according to oil removal situation. Oil-skimming net can also be effective to prevent oil contamination. Oil-floating zone can be reduced by simultaneous action of oil removal by the adsorptive sheet attached to net and the enclosure operation of net.

2.2.4.2. Separation and removal of oil from water. This section concerns removing small amount of water from oily materials, or small amount of oil from water. Figure 5 is a typical example of the system for the removal [7]. The coalescer cartridge medium is composed of microfiber nonwoven felt having very small pore size. Small size of oil particles can quickly coagulate by the microfiber network as schematically shown in the figure (a). The separator cartridge is made of membrane, screen or nonwoven having surface of water repellency. This kind of system is used for such end uses as the removal of water from aircraft fuel, the removal of water from petroleum products, and the removal of oil from cooling water for chemical plants.

2.3 Air purification

2.3.1 Bag filter

Bag filters are widely used for cleaning the exhaust gas from several kinds of incinerators. They are usually made of glass fiber-woven fabric or synthetic fiber needle felt or its combination with woven fabric. Polyphenylene sulfide, *m*-aramid, polyimide, and polytetrafluorocarbon are typically used as the synthetic fiber material. A typical example of bag filter system is shown in Figure 6. The dust contained in exhaust gas is accumulated on the surface of a bag filter whose shape is tubular. When the inlet pressure increases up to a settled value by the layer-up of accumulated dust, it is dropped off from the bag surface by counterpulse jet or mechanical vibration. Acid gas and some other harmful gas substances can also be removed by introducing such materials as slaked lime and activated carbon in the bag.

2.3.2 Air filter

Air filters used for the removal of dust for clean room and office contribute to energy saving by the recycling of conditioned air. The filter medium is usually nonwoven. With a decrease in the fiber diameter, the removal efficiency of filter is increased. But its pressure

(a)

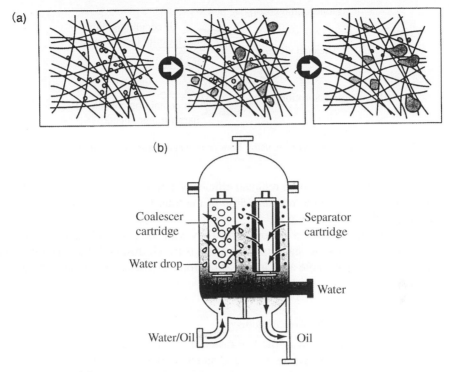

(a) principle of separation at coalescer medium

(b) separator tank with coalescer and separator cartridges

Figure 5. Oil/water separator [7]: (a) principle of separation at coalescer medium and (b) separator tank with coalescer and separator cartridges.

drop is also increased. Removal efficiency is the lowest at 0.05–0.1 μm of dust particle size, because inertial collision effect for catching dusts within airflow by fiber within the medium becomes more significant for larger particles with particle size above 0.1 μm, and diffusion effect is more effective for small particles, below 0.05 μm. Filter medium is

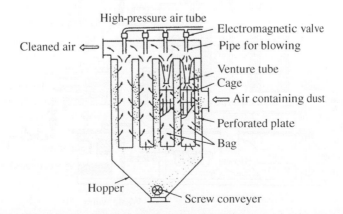

Figure 6. Schematic illustration of a bag filter system [8].

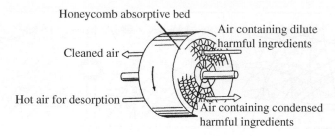

Figure 7. Removal system for toxic gas using honeycomb bed made of activated carbon paper [9].

usually pleated in the filter of high removal efficiency, because pleating increases the area of filtration and can reduce the pressure drop of the filter by lowering the velocity of air passing through the filter medium.

Electret filter can have higher removal efficiency with comparatively low pressure drop by making the use of electric static traction force between dust and fiber. The filter unit whose medium is exchangeable after its pressure drop increases up to the settled value has been developed in order to save unit frame.

2.3.3 Toxic-gas removal and solvent recovery

Activated carbon fiber is useful as the key material for both the removal system of toxic gas and the solvent recovery system. Figures 7 and 8 are typical examples of system using activated carbon fiber for harmful exhaust gas and for solvent recovery, respectively [9].

In the former system, the honeycomb bed for adsorption is formed by piling corrugated board made of activated carbon paper. The exhaust gas containing harmful substances passes through the main part of the rotating honeycomb, and the harmful substances are removed by the adsorption at the honeycomb wall. In the fixed zone of smaller angle, they are desorbed by hot air. The air containing condensed harmful substances thus produced is re-treated by much-smaller-scale system of oxidization or of adsorption for solvent recovery. This system is feasible for treating exhaust gases from several kinds of industries such

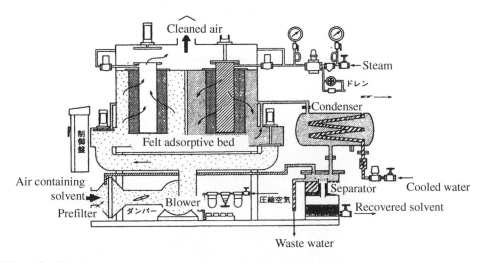

Figure 8. Solvent recovery system using activated carbon fiber felt [9].

as semiconductor production, liquid crystal display production, and automotive painting, which generates large amounts of airflow containing small amounts of harmful substances.

In the solvent recovery system, activated-carbon-fiber nonwoven is wound on a plural of cylindrical cages. Solvent adsorbed by the cylindrical bed is periodically desorbed by steaming. Thus solvent removal can be performed with continuous operation. The advantages of this system include the higher quality of recovered solvent and the smaller size of system over systems using conventional granular activated carbon, because of much higher absorbing and desorbing rate of activated carbon fiber.

2.4 Contribution to energy and resources matters

2.4.1 Fiber materials for battery and for fuel cell

2.4.1.1. For electrode. The electrode of fuel cell must have high gas permeability with high electric conductivity. Carbon fiber sheet is useful as the material for this electrode. Activated carbon fiber is also promising as the electrode material of battery for the storage of night electric power. Carbon nanofiber and nanotube are very useful as electrode material for several kinds of battery because of their good electric conductivity and high specific surface area.

2.4.1.2. For separator. Wet-formed nonwoven made of such fiber material as PP is used for the separation sheet of nickel/hydrogen battery. The sheet is composed of fine fibers and is finished by hydrophilic treatment, which enhances its working life, its electric power and its ability to suppress self-discharge.

2.4.2 Composites for high-pressure vessel, wind turbine blade, and the rig of deep-water oil plant

2.4.2.1. High-pressure vessel. Compressed natural gas (CNG) is more environmentally friendly than gasoline because CNG emits less CO_2 than gasoline. Its vessel can contain the CNG of 250 atm pressure. The CNG vessel made of carbon fiber reinforced plastic (CFRP) is 70% lighter than that of steel. The former is generally fabricated by filament winding on Al liner.

Highly pressured hydrogen is one of the most realistic candidates as the fuel source for the fuel cell of automobile. Its pressure is required to be about 700 atm. Hence CFRP vessel using high-performance carbon fiber is indispensable to keep the weight of vessel lighter than 50 kg/unit.

2.4.2.2. Wind turbine blade. The total amount of electric power produced by the installed aerogenerators has been growing rapidly. Most of their turbine blades are made of glass fiber reinforced plastic (GFRP). But with an increase in their size in order to attain higher efficiency, CFRP becomes a more suitable material for their blade. It is usually fabricated by vacuum-assisted resin-transfer molding [10] method using multiaxially layered warp knit [11].

2.4.2.3. Rig tools of deep-water oil plant. In order to cope with the growing demand of oil and saturation tendency of its production amount, the necessity of oil digging in the area of deep-sea water has been more serious. The tether and riser made of steel for such a rig

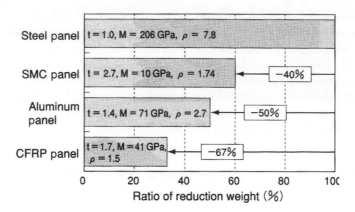

Figure 9. Comparison of weight reduction effect by panels made of several materials under the condition of same stiffness [12]; SMC indicates Sheet molding compound; CFRP, carbon fiber-reinforced plastic.

cannot resist their own heavy weight. Then CFRP composites have become indispensable material for them.

2.5 Contribution to energy saving

2.5.1 Lightening transport vehicles

As suggested in Figure 9, changes from metal to GFRP and CFRP composites can reduce the weight of transport vehicles. The weight reduction can be connected to energy saving.

In the case of passenger aircraft, use ratio of CFRP has gradually increased. Figure 10 schematically shows the materials used for new aircraft B787. The use ratio of CFRP in the weight base of B787 has reached about 50%, and some of its parts are made of GFRP.

In the case of automobiles, many metallic parts such as bumper beam, engine cover, battery bracket, front end, underbody shield, and battery tray have been objective parts to be replaced by GFRP. Figure 11 shows some examples of such replacements. In the case (a), the intake-manifold is fabricated by injection molding and ultrasonic adhesion. Its weight

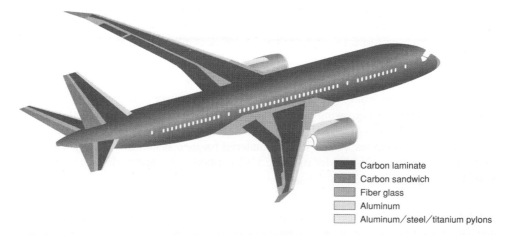

Figure 10. Materials used for Boeing B787 [13].

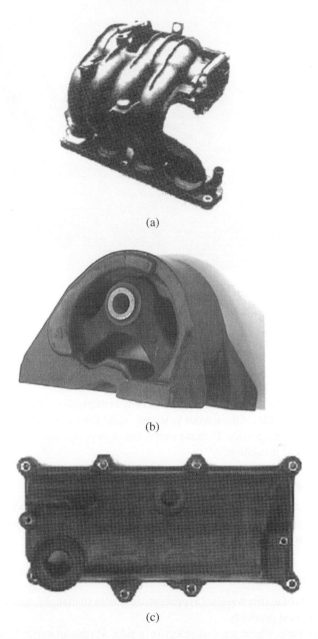

(a)

(b)

(c)

Figure 11. Examples of automobile parts made of glass fiber reinforced plastic: (a) Intake-manifold made of PA6/GF [14], (b) Mount bracket made of PA66/GF [15], and (c) Engine head cover made of PA66/GF [15].

reduction ratio is 40% from previous aluminum part. In the case (b), the mount bracket is fabricated by injection molding, and its weight reduction rate is 50%. On the other hand, the use of CFRP parts to passenger cars has been fairly limited. Figure 12 is an example of the advanced cars in which many parts are composed of CFRP. In this case, most of the exterior panel, most of the chassis, and the shock absorber are made of CFRP. Its use of carbon fiber is more than 110 kg/unit. In the near future, introduction of CFRP into

Figure 12. SLR-Maclaren (CFRP richly used car) [16].

automobile parts and frame will be gradually increased with an increase in the demand of energy saving.

2.5.2 Thermal-insulation fibrous materials

We can mention glass wool, rock wool, and fiber fill as thermal-insulation fibrous materials related to technical textiles. Glass wool is widely used for such end uses of insulation as house, building, and refrigerator. In these cases, the high performance of air-stagnant layer is utilized for thermal insulation.

2.6 Recycling and reuse in terms of technical textiles

In this section, recycling of textile wastes into technical textile products is discussed. The former corresponds to defensive technologies and the latter to offensive technologies through material saving.

2.6.1 Recycling and reuse from textile wastes

There are two kinds of textile wastes: (a) processing wastes in manufacturing textile products and (b) wastes of used products.

In the former case where we can selectively take fibrous material from these wastes, recycling can be effectively carried out, because it does not usually contain so many different kinds of fibrous materials. Wastes consisted of PET fiber can be chemically recycled [17]. Fibrous material of nylon 6 taken from such products as carpet can also be chemically recycled [18]. Nylon 66 fabric taken from airbag can be reused as a main material for engineering plastic [19].

But in case the fibrous material is firmly combined to other kinds of material, recycling is not easy. In the case of CFRP wastes whose matrix is made of thermoset resin, the resin is removed after its chemical degradation and then the remaining carbon fiber can be reused as valuable reinforcing material of short fiber. Waste of GFRP is usually broken into small particles which are then used as filler of cement etc. Reuse of general shredder dust from automobile is extremely difficult. But the dust sorted as mixture of fibrous materials and urethane has been transformed as automobile silencer boards settled between engine, cabin, and under-floor carpet [20].

The automobile parts made of technical textiles are necessarily reused in the reused car. This is one of the most popular examples of reuses of technical textile products.

2.6.2 Recycling of fibrous materials into technical textile products

Recycled fibrous materials have been widely used for cushion material and wiping sheet. In addition to these applications, several trials for recycling into technical textile products have been conducted. We can mention such applications to man-made wood, floorboard, substrate sheet for cultivation, reinforcement of earth, and silencer as mentioned above. An example of a boat made of the man-made wood produced from recycled fibrous material has been reported. One of the other examples of recycling from textile mills is the application to substrate sheets for lawn cultivation.

2.7 Water-saving system for plantation

There is a large area of land in the world where water resource is too insufficient to make and to keep plantation. Therefore it is thought that the efficient use of water for plantation is very important to be developed. Nonwoven has been widely used for controlling sunlight by covering agricultural plant. But the use of fibrous materials for irrigation and water retention for plantation has been very limited. But there have been several remarkable proposals of irrigation system for water saving.

3. Textiles products for automobiles

3.1 General scope of textile products in automobile sector

The automobile sector is the largest user as estimated from Table 1. The amount of fiber used for a standard passenger car is about 25 kg. This is now increasing because the requirements for the safety and comfort of passengers and for car weight reduction, which means an increase of reinforcing fibers for hard composites, are being intensified. Furthermore, the worldwide demand for automobiles is growing. Hence, the total amount of automotive fiber is increasing. These facts indicate the industrial importance of automotive fiber in terms of end use.

The selection criterion for automotive-use materials is "performance/cost" similar to most other technical textile materials. But it is generally more strictly applied to automotive use. For example, steel cord is usually used as a reinforcing material in tyre-belt plies for passenger cars. The reason is that steel has the largest value for "tensile modulus/cost" of the existing materials. But if car weight reduction becomes more important in the future, the selection criterion may become "specific tensile modulus/cost" and then steel could be replaced by a high-performance fiber such as polyketone fiber [21].

Fiber is used in the manufacture of several kinds of parts in automobiles such as (a) tyres, driving belts, tubes, and hoses; (b) seat belts and air bags; (c) seats, roof trims, and floor coverings; (d) noise control materials; (e) cover sheets; (f) filters; and (g) mechanical parts, exterior body panels, and bumper beams. In the near future, optical fiber for the information and driving-control system and reinforcing fiber for the fuel pressure vessel may also be widely used. The percentage use of automotive fibers for main parts is summarized in Table 2. "Hard composites" is the second largest user group, as shown in the table. They are reinforced mostly by glass fiber through compression molding and/or injection molding.

With regard to the shares in the total amount of automotive fibers used, PET fiber takes up about 42% and nylon 66 fiber about 26% [22].

Table 2.   The amount of automotive fibers by end use share [22].

End uses (main parts)	Share (%)
Rubber composites (tire, driving belt)	36
Safety system (seat belt, airbag)	4
Car interiors (seat, door trim, roof trim, floor covering)	17
Hard composites (body, mechanical parts)	30
Others (filter etc.)	13

3.2 Rubber composite parts

3.2.1 Tyres

Tyre structures can be classified as bias tyres and radial tyres. These are illustrated in Figure 13. Bias tyre has a higher energy-absorption capacity; therefore, it is more feasible to use it on rough roads. But it also has a lower wearing resistance. Hence it has been gradually replaced by radial tyre with the increase of smoother road surfaces. In bias tyre, nylon fiber

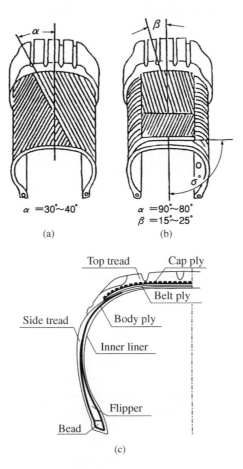

Figure 13. Structures of tyres for passenger car [23]: (a) bias tyre, (b) radial tyre, and (c) the cross-sectional structure of radial tyre.

Table 3. Properties of nylon 66 and HMLS PET tyre cords comparing with other materials [24].

	Nylon 66	High-modulus low-shrinkage polyethylene terephthalate	Polyethylene naphthalate (PEN)	Rayon super III	p-Aramid	Steel
Thickness	1400/2	1670/2	1670/2	1840/2	1670/2	–
Twist (T/10 cm)	39 × 39	39 × 39	39 × 39	47 × 47	32 × 32	–
Strength index of flat yarn	100	90	80	55	210	–
Elongation of flat yarn (%)	19	11	9	9	3.6	–
Thermal shrinkage of flat yarn (%)	5.5	3.0	1.5	≥ 0.3	≥ 0.3	–
Elongation at fixed load (%)	7.0	4.0	2.0	2.0	0.9	–
Creep index	100	50	–	70	10	–
Adhesiveness	Good	Fairly good	Fairly good	Good	Fairly good	Good
Specific gravity	1.14	1.38	1.36	1.52	1.44	7.81
Melting point (°C)	265	260	272	Degrade	Degrade	1450
Glass transition temperature (°C)	–	70	113	–	–	–

Note: Rayon super III indicates the third-generation super-tenacity rayon.

is usually used for body ply cord because of its excellent toughness. In radial tyres for passenger cars, PET fiber is mainly used for body ply cord because its higher modulus can contribute to improved driving comfortability by the reduction of flat spotting. Nylon 66 fiber is mainly used for cap ply cord because of its higher strength and higher toughness.

A high-modulus low-shrinkage type of PET multifilament is usually used for tyre cord because higher modulus and higher heat resistance in shrinkage are important requirements of tyre cord. But an important problem that needs to be improved for PET is its comparatively low adhesiveness to matrix rubber. Regarding nylon, a high-strength type of multifilament is required for tyre cord. The properties of tyre cords made of several materials are summarized in Table 3 in which yarn thickness/number of plies is also shown.

There are some other methods for the fiber reinforcement of tyres than using cord. One is a direct blend of short fiber such as p-aramid with matrix rubber. Another example is the use of the composite made of p-aramid nonwoven and rubber, which is inserted into the inner part of a truck/bus tyre to increase run-flat capability.

3.2.2 Driving belts

Driving belts can be generally classified as V-belts, V-ribbed belts, cogged belts, and metal-combined belts, as shown in Figure 14. For subsidiary automotive equipment such as air conditioners, V-belt is used. Cogged belts and V-ribbed belts are used for engines. Recently, a metal-combined belt has been used for speed variation in small passenger cars. PET fiber is applied to the cord of V-belt and the cloth of its upper part. p-Aramid fiber is applied to the cords of V-belts, V-ribbed belts, and metal-combined belts.

3.3 Passive-safety inner parts for automobiles

Seat belts and air bags are very important as passive-safety systems for automobiles.

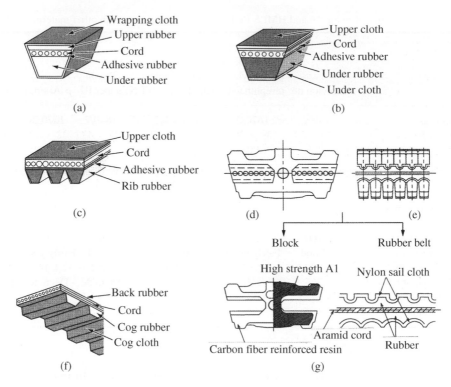

Figure 14. Structures of several kinds of driving belts [24]: (a) wrapped V-belt, (b) raw edge V-belt, (c) V-ribbed belt, (d) block, (e) rubber belt (f) belt with cog, and (g) composite V-belt.

3.3.1 Seat belts

A seat belt is used to prevent a passenger in a car from being thrown off his or her seat in an accident by fixing him or her to the seat and absorbing the impact shock. PET woven fabrics are usually used for their webbing. PET is more suitable than nylon for the webbing because PET has a higher impact-energy-absorbing capability and suffers less discoloration by sunlight. There are several standards for the webbing such as the Federal Motor Safety Standard, Economic Commission for Europe and Japan Industrial Standard. The main mechanical requirements of these standards are strength, width, elongation, and energy absorption ratio. There are also requirements related to durability such as wearing resistance, cold and heat resistance, water resistance, light-degradation resistance, and color fastness. Typical examples of specification values of PET fiber for this end use are yarn thickness 1000–2500 dtex with 4–25 dtex of its monofilament thickness and strength 8–10 cN/dtex.

3.3.2 Air bags

The function of an air bag is to cushion a passenger from collision impact. The bag is instantly inflated by heated gas after a collision stronger than a certain limit. Air bags can be very effective as a passive-safety system in combination with a seat belt. There are several kinds of air bags – driver bag, front passenger bag, thorax bag, curtain bag, rear bag, and knee bag. Figure 15 shows photographs of several kinds of air bag inflated.

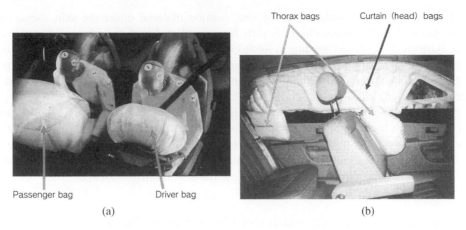

Figure 15. Examples of inflated airbags [26]: (a) front airbags, (b) thorax bags and curtain bag.

The base material of an air bag is nylon 66 weave because nylon fiber can be more compactly contained and has higher resistance to small burning particles. The thickness of nylon filament used for air bags ranges from 235 to 700 dtex. The number of monofilaments ranges from 70 to 220. A high-strength type is needed. There are two types of base material – coated and noncoated. The coating material is now mainly silicone resin. The advantages of the coated type are better nongas permeation, easier bag-pressure control, and greater heat resistance to burning particles. On the other hand, the noncoated type is lighter in weight, thinner, more flexible, and less expensive. For noncoated cloth, some special products have been developed by weaving with an ultra-high yarn density, with a filament of lower monofilament thickness, and with a filament whose fiber cross section is flat.

3.4 Car interiors

The main interior components of an automobile are the seats, door trim, roof trim, and floor covering.

3.4.1 Seats and door trim

Polyester fiber is used for most of the seat skin sheet and for some seat cushion material. It is also used for some door trim skin material because PET fiber has a higher modulus, higher heat stability, higher resistance to color change, and higher durability for sunlight degradation and is less expensive.

There are several kinds of seat skin sheets such as pile weave, weave, tricot with raising, pile double raschel knit, and pile circular knit. General trends are toward an increase in knit fabrics (tricot, double raschel, and circular knit) that are less expensive and have more formability than weave fabrics. Recently, a suede fabric, using a PET microfiber nonwoven as base material, has been introduced. Skin sheets containing phase-change material (PCM) have also been developed, which can increase the microclimate comfortability of the seat. The essential properties of seat skin sheets can be classified as aesthetic effect, physiological comfortability, strength/wearing durability, color fastness, flame retardancy, heat resistance, and nonvolatile substance content. They are also often required to have some special functions such as antibacterial, deodorizability, antistatic and stain-resistant properties. Pile or raising of these sheet fabrics is usually related to an increase in the values of tactile and visual aesthetic effects. An extremely high level of color fastness in sunlight is usually required because automobiles can be used in the environment of high temperatures and strong sunlight.

Urethane foam is usually used for seat cushion material under the skin sheet. But some special PET nonwovens have recently been developed such as a fiber mass stabilized with elastomer fiber-fused bonding, a folded web and stitch-bonded web, and a PET three-dimensional knit fabric with super water absorbancy. One of the most important advantages of these fibrous materials over urethane foam is their good moisture permeability, which keeps passenger's physiological comfort.

The material of door trim skins is usually made of plastic such as vinyl chloride and polyolefin. But textile sheets are also used for higher-class cars. In most cases, the textile material is the same as the seat skin fabric. However, the fabric needs to have high enough formability to be made into the complicated shape of a door trim. Its lower end is usually covered by the same carpet as the floor because the door often gets kicked.

3.4.2 Roof trims

PET nonwoven and tricot are used as roof trim skin sheets. The use of needle-punched nonwoven in particular has increased, and nonwoven patterned using velour by needling is especially common. But there are also spun-laced nonwovens and stitch-bonded nonwovens. They are usually formed into roof trim by integration with base materials. Pigment-dyed PET has been widely adopted. Sheets need to have color fastness for sunlight, heat resistance, mechanical durability, light durability, formability, and nonvolatile substance content and stain resistance in addition to lightness. Certain levels of sound-absorbing capability and heat insulation are also needed.

The base materials for roof trim can be mainly classified as polymer foam sheets and fiber-reinforced porous polymer sheets. Glass fiber is usually used for reinforcing these sheets.

3.4.3 Floor coverings

Floor coverings can be divided mainly into liner coverings and optional coverings. Needle-punched nonwoven carpet and tufted carpet are used for liner coverings. The usage of nonwoven carpet has significantly increased because it is more economical and has better formability. There are three main kinds of surface patterns – plain, loop like, and velour like, as schematically shown in Figure 16. The latter two patterns can be made by the combination of specific needles and needling patterns. The surface material of these needle-punched

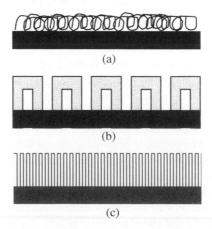

(a)

(b)

(c)

Figure 16.   The illustration of three main types of surface patterns for needle punched carpet [27]: (a) plain, (b) loop like, and (c) velour like.

carpets consists mainly of pigmented PET fiber and/or pigmented PET recovered from bottles. Its fiber thickness is reduced from 11 to 6 dtex to increase its covering factor. Quite attractive aesthetic effects can be achieved with needle-punched carpet by the specific patterning described above. The main performance requirements are mechanical durability, sunlight durability, heat resistance, sunlight color fastness, sound absorbability, sound insulation, antifogging, nonvolatile substance content, flame retardancy, stain resistance, and formability. With an increased requirement for weight reduction, sound absorbability has become more important than sound insulation.

Tufted carpet has usually been used for optional carpets, where its aesthetic effect and appearance of high quality to the customer is of greater importance. Its weight ranges from 350 to 2000 g/m^2 on the basis of the wide variety of customer requirements. Its surface material fiber is nylon, PET, or PP. PP is now increasing its usage share.

3.5 Filters for automobiles

Automotive filters can be divided as engine filter, cabin filter, oil filter, and filter for diesel engine exhaust gas.

3.5.1 Engine filter

Its function is to remove the dust of such materials as sand and carbon particles, which are contained in the air taken into automotive engine. There are two kinds of filter materials – paper and nonwoven. The paper is usually made of wood pulp and such a material as cotton linter or man-made fiber. Its main advantage is high removal efficiency. It is used in highly pleated form. But its use life is still comparatively short because its filtration is carried out only at its surface. In the case of nonwoven, filters having fiber thickness gradient and density gradient structure have been developed. They can effectively remove both coarse and fine dust particles by bulk filtration.

3.5.2 Oil filter

It is used for purifying engine-lubricating oil. The objective particles to be removed by the filter are carbonized fine particles grown by the heat of the engine and metallic particles/sludge contained in the oil. Its filter material is paper made of such materials as cotton linter, wood pulp, rayon, and PET fiber or nonwoven that is formed by wet process and then finished by phenol resin. In the case of filter used for diesel engine, specially high removal efficiency for carbonized fine particle is required. Highly fibrillated man-made fiber has been applied to this kind of filter.

3.5.3 Cabin filter

Removal of the dust that comes from the outside and the inside of cabin and deodorization of such smell as smoking are the purposes of settling cabin filter. In order to remove fine particles from exhaust gas from engine and fine mists from smoking, electret filter is conveniently utilized. Deodorization can be done using sheet sustaining activated carbon particles chemically modified.

3.5.4 Filter for diesel engine exhaust gas

Figure 17 shows a system for removing particles contained in the exhaust gas from diesel engine using felt made of SiC group fiber [33]. The felt is pleated into cylindrical forms. Particles filtrated by the filter are burned out at times by electrical heating. Its removal efficiency is more than 95%.

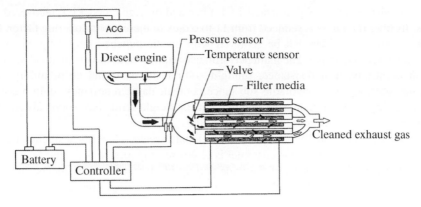

Figure 17. System for removal of particles exhausted from automotive diesel engine [32]; ACG indicates alternating current generator.

3.6 Noise control materials for automobiles

The sound insulation capability of a sheet is generally proportional to its weight. For example, 10 kg per car is needed for a good-sound-insulation floor covering. In order to reduce car weight, the strategy for noise reduction within the cabin has been changed to adopting the use of parts having higher sound absorption capability instead of higher sound insulation. Interior parts such as floor covering and roof trim are typical targets in this strategy, as stated above. Multilayered nonwovens are used for the engine silencer and sheet supporting dumper to reduce unusual sounds in line with this strategy. Figure 18 is an example of such an engine silencer.

4. Textile products for medical and hygienic use

4.1 Medical uses

4.1.1 General scope of textiles in medical/biological uses

Requirement for higher level of medical treatment and higher quality of life has been increasing the importance of applications of textiles to medical uses. In addition, medical science based on biology is now rapidly progressing. Therefore, there is still a big space where fiber and textile technologies in this field can further be effectively utilized.

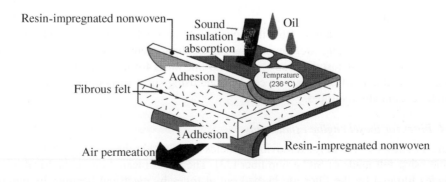

Figure 18. An example of engine silencer [34].

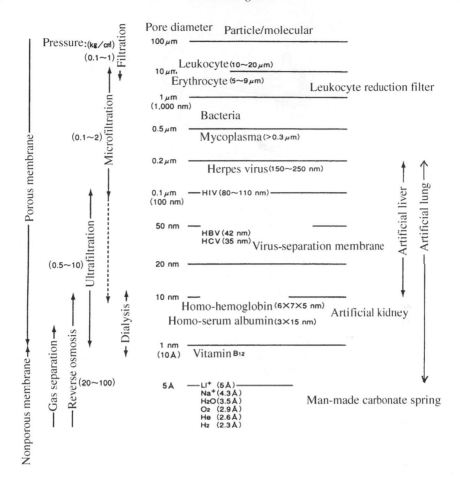

Figure 19. Several kinds of separation functional fibrous products for medical use in terms of pore size [35].

There are many kinds of fibers and textiles that have their own separating functions for medical use, as shown in Figure 19. These functions are originated from the action of sieving molecular/particle through the pore or molecular spacing of their own specified size.

In the first part of this section, membrane hollow fibers for artificial kidney, artificial lung, and some other medical treatments are introduced. Then a system using fibrous material for leukocyte reduction, which is also medically used, and some applications of fiber/textiles to tissue/biological engineering will be introduced. Finally, miscellaneous applications of fibrous materials to medical and paramedical uses are described.

4.1.2 Membrane hollow fibers for artificial kidney, artificial lung, virus removal and others

4.1.2.1. Artificial kidney. With a fall in the functional ability of kidney, water and the waste substances such as urea that should be discharged as urine gets accumulated in blood, causing uremia.

Artificial kidney is the system that artificially purifies the blood on the basis of dialysis. Its module is consisted of a large number of hollow membrane fibers, as shown in

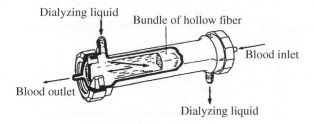

Figure 20.   Artificial kidney module containing a bundle of hollow fibers.

Figure 20. Water and waste substances contained in blood are introduced into the hollow part and diffuse through fiber membrane on the basis of their concentration gradient. The typical example of inner diameter and membrane thickness of the fiber is 200 and 10 μm, respectively. Several kinds of materials for the membrane have been developed such as regenerated cellulose, triacetate, polyacrylonitrile, polymethylmethacylate, polyvinyl alcohol, and polysulfone. The material is selected and/or modified in order to avoid bioincompatibility troubles such as complement activation and/or leukocyte reduction. Several kinds of countermeasures to realize the uniform contact of dialyzing liquid with membrane fibers have been taken. They include uniform fiber crimping, spiral winding of thin yarn on fiber, and the introduction of a certain projection to fiber.

Human kidneys have sharp sieving characteristics, as shown in Figure 21. Sieving characteristics of previous conventional artificial kidneys was so much different from that of a human kidney that the accumulation of β2-microglobulin has caused the problem of serious symptom. Thereafter, the improvement of artificial kidneys in sieving characteristics toward human ones has been carried out.

4.1.2.2. Artificial lung. Artificial lung is used for backing up heart surgery. Hollow fibers whose pore size is less than 1 μm are a key component in the lung system. Pressured oxygen and carbon dioxide-rich blood are introduced into the inside and the outside space of the hollow fiber, respectively. Gas exchange between oxygen and carbon dioxide is carried out

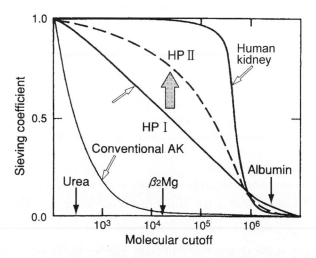

Figure 21.   Sieving characteristics of artificial kidney and human kidney [36].

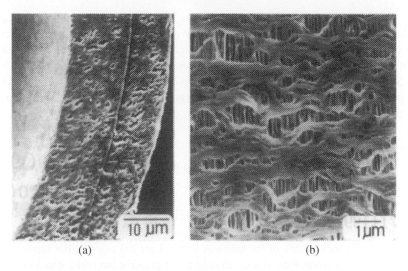

(a)   (b)

Figure 22. Enlarged view of membrane hollow fiber used for artificial lung [37]: (a) the cross-sectional view of the membrane where an active layer is located in the center and (b) the view of the outer surface.

by diffusion through the membrane. Figure 22 shows an example of hollow fiber used for the system, whose gas exchange rate is extremely high because of a very thin active layer in the middle. The active layer consists of amorphous polyurethane and is supported by the two outer porous layers made of high-density polyethylene (PE).

4.1.2.3. Virus removal filter. The filter must have both sufficient permeability for plasma protein whose range is from about 1 to 10 nm and sufficient removal rate for virus including HIV whose size ranges from about 10 to 100 nm. The material of this UF membrane hollow fiber is cellulose. A kind of specific capillary-void structure is extended within the membrane in its thickness direction, whose size is strictly controlled to satisfy the above condition.

Figure 23 shows an enlarged cross section of the membrane, which is capturing HIV. It is said that this method is excellent in easy use, nondegradation of blood and removal reliability than the other method of virus removal methods such as heating and low-PH treatment.

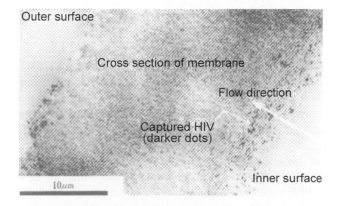

Figure 23. Cross-section view of virus removal filter and captured HIV [35].

4.1.2.4. Artificial liver. Artificial liver is expected to substitute the human liver up to the time when it will restore its functionality. The liver is a kind of bioreactor in which many biochemical reactions take place, making it almost impossible to artificially realize its function on the basis of these biological reactions. Hence, in the artificial liver that has been recently developed, a liver taken from an animal such as a pig is introduced into the hollow part of fibers to make a cylindrical cell organ. The blood or plasma of the patient is circulated in the outside space of the fibers.

4.1.3 Filter for leukocyte removal

Leukocytes, whose role is mainly to attack foreign substances invading the human body, are contained within 1% in the blood. But this function often causes several side effects in blood transfusion.

A system using special PET nonwoven has been developed, which can remove leukocytes with the efficiency of more than 99.99% while retaining the sufficient number of thrombocytes. The key points of the nonwoven used are (a) use of fiber having an ultrasmall diameter, (b) introduction of a small amount of positive ion residue to a coating resin on fiber surface, and (c) selection of suitable blood flow rate through the filter.

This filter is widely used for the prevention of side effects at blood transfusion in the world. Another recent application to be noted is leukocytapheresis for the ulcer of the large intestine.

4.1.4 Application of tissue engineering

4.1.4.1. Artificial blood vessel. The structural material of an artificial blood vessel of middle and large diameter is usually PET fabric or polytetrafluoroethylene membrane. The tubular vessel is corrugated like bellows, to be flexible. In the case of PET fabrics, some of them have velour surface and/or are made of microfiber. Tissue culturing to such a tube is easier because cell has a tendency to be more easily cultured along fine fiber. To avoid the leakage of blood, some vessels have usually a barrier layer made by filling such materials as gelatin or made of a biocompatible elastic polymer.

4.1.5 Miscellaneous fibrous products for medical and paramedical uses

We can mention endotoxin removal module, wound dressing goods, surgical thread, artificial ligament, and medical plaster as the other medical goods using fibrous products. Gauze/towels, surgical gown, surgical drape, scrub suit, surgical cap, masks, sheets, paramedical plasters, etc., can be mentioned as fibrous products for paramedical use. Many surgical goods are disposable and are required to be lint free and to have bacteria barrier capability. Nonwoven made by fluid jet entanglement [38] and melt-blown [38] web reinforced by spun-laid [38] web such as SMS (nonwoven made by the process which is combined of spunlaying (S), melt blowing (M) and spunlaying (S)) are major materials for these goods. Some goods must be water repellent and some other goods must be water absorptive. In this section, three kinds of goods based on a noticeable technology are introduced.

4.1.5.1. Fibrous absorbent for the treatment of septicemia. The absorbent fiber can selectively remove the endotoxin from blood, which is the cause material of septicemia. It is a bicomponent fiber of sea-and-island type whose sheath material is polystyrene. Polymyxin B, which is a polypeptide antibionic, is chemically fixed to the porous styrene sheath of the fiber, as shown in Figure 24. The removal is carried out on the basis of the chemical affinity

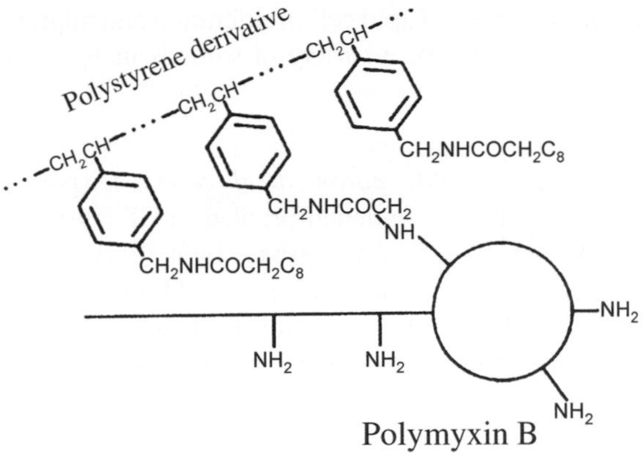

Polystyrene derivative

···-CH₂CH- ···-CH₂CH- ···-CH₂CH- ···

CH₂NHCOCH₂C₈

CH₂NHCOCH₂
NH

CH₂NHCOCH₂C₈

NH₂

NH₂   NH₂   NH₂

NH₂

Polymyxin B

Figure 24. Fixation mechanism of polymyxin B to fiber [39].

between the endotoxin and polymyxin B. The knitted fabric made of the adsorbent fiber is wound on a tube and contained in the module, as shown in Figure 25. The circulating blood in the treatment system is purified while passing through the module.

4.1.5.2. Products for wound dressing. Several kinds of materials for wound dressing have been developed. But in this part, the following three examples are introduced as their representatives.

Chitin fiber made from crab shell is widely used for wound covering/artificial skin carrier in the form of nonwoven/paper and sponge. It is dissolved into human body by human enzyme and effectively enhances the healing of wounds. These wounds include burns/scalds, bedsores, ulcers, and several kinds of external injuries. It is also used as the carrier for cultured skin [40].

Partially carboxymethylated cellulose fiber has high absorbencies of both water and saline. Then cotton gauze modified with the carboxymethylation used for wound dressing can take up wound exudates. Cohesive gel sheet obtained by carboxymethyl treatment on solvent-spun cellulose fiber has been widely used in wound management [41].

Combined sheet called as Biobrane, in which perforated silicone rubber and nylon fabric are bonded by collagen-peptide gelatin, has been developed. The gelatin is biocompatible

Figure 25. Module containing adsorbent fiber knit for sepsis [39].

and can ease the pain of the patient. Rapid cell proliferation and migration into nylon fabric occur. It can lead to faster healing of a variety of wounds including full-thickness burns [42].

4.1.5.3. Fiber material for reusable paramedical goods. Disposal of surgical goods through incineration causes an environmental problem. PET fabrics are rather inferior in reusability because PET tends to be easily degraded by hydrolysis by autoclave treatment for sterilization. Higher temperature treatment is required for higher sterilization efficiency. A partially aromatic polyamide fiber that is feasible to such high-temperature treatment has been developed as fiber material for reusable paramedical goods [43].

4.2 Hygienic uses

In this section, sanitary napkins and diapers are shortly explained as representative goods for hygienic uses. The amount of nonwoven used for such hygienic goods is the largest in share among all the application fields. In such developing countries as China, they are now rapidly growing.

4.2.1 Sanitary napkin

Sanitary napkins are usually composed of three layers. Their inner surface layer that contacts the skin is made of nonwoven. The performance items required are (a) the quick transfer of menstrual blood from the inner surface to the central layer and the stop of counter blood flow to the inner surface, (b) to give feeling of comfortable softness and pliability, and (c) wet mechanical sustainability. Its central layer composed of super absorbent polymer absorbs the transferred blood. Its outer surface layer is usually made of polyethylene (PE) film to guard the leakage of the blood.

The nonwoven is usually made by spun-laying [38] or SMS (see section 4.1.5 in the former book [1]) combined with thermal bonding [38] or with water jet entanglement [38]. Some nonwovens have pore size gradient for better efficiency of blood transfer by the enhancement of capillary effect.

4.2.2 Diaper

Diapers can be classified as pant type and pad type according to their form, and as baby use and aged person use according to user. But the layer components are almost the same among all these diapers and are similar to those of sanitary napkins. In some cases, nonwovens having fairly high stretchability are used.

5. Textile products for protection and safety

5.1 General scope of textiles in protection and safety

Strengthening of the belief of humanism and increase in several kinds of external physical threats to human life and activities have made protective/safety textile goods more important. Hence their market demand is steadily growing. Higher level of functional performance has also been required from the goods.

There are such typical external threats as natural disaster, traffic accident, war, fire, criminal offense, working accident, and leisure/sport accident. Textiles of protection/safety are defined as textiles used for the goods that can personally avoid, guard, or reduce the damage caused by external threats. But the performances required from these goods and their product structures to realize the performances are fully dependent on their own objective of protection.

In this section, textiles protective from bullet/stab disaster, danger of fire fighting, and several harmful environmental offenses, etc., are introduced.

5.2 Bulletproof and stabproof

5.2.1 Bulletproof

Bulletproof jackets usually take the form of panel structure piled by woven fabrics with high yarn density, which are made of organic fibers that have mechanically high performance properties, such as *p*-aramid, UHMW-PE, and *p*-phenylene-2,6-benzobisoxazole fibers (see sections 3.4, 3.5, and 3.7.1 in the former article [1]). The kinetic energy of the bullet is dissipated by the fabric into (a) the deformation of the panel, (b) the breaks and the deformation of fibers, and (c) the deformation of the bullet. The performance of a bulletproof textile depends positively on the aerial weight of the fabrics and the breaking energy of the fiber, and negatively on the single-filament thickness.

There is another kind of panel for bulletproof textile whose manufacturing method and structure are schematically shown in Figure 26. In this panel, fibers are straightly aligned and can carry the impact energy of the bullet more effectively than the fibers in the woven fabric.

5.2.2 Stabproof

In the stab resistance, the key for its countermeasure is how effectively it makes the apex point round by hard material. One example of panel structures is composed of woven fabric with high yarn density, which is made of UHNW-PE fiber and glass fiber. Another example is composed of *p*-aramid-fiber-woven fabric covered by ground-stone particles [44].

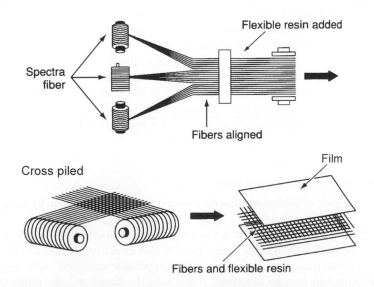

Figure 26. Manufacturing method and partial structure of a panel for bulletproof [44].

5.3 Suit for fire fighting

Suits for fire fighting are usually composed of three layers – outer surface layer, central layer which functions as breathable waterproof, and inner layer whose function is thermal insulation. m-Aramid, p-aramid, p-phenylene-2,6-benzobisoxazole, polyamide-imide (see sections 4.3 and 3.7.1 in the former article), or flame-retardant finished-wool are examples of fiber materials used for these layers. For breathable waterproof layer [45], coating or laminating is applied to the central-layer fabric. Thermal insulation is made by the effective use of air layer that is contained in the inner layer [46].

5.4 Chemicals protection and nuclear particles, biologically toxic materials and chemically toxic materials (NBC) protection

The chemical protective clothing functions as a barrier from outer harmful chemicals, which may be liquid, gas, mist, or particles. There are several kinds of clothing belonging to this category, such as those of industrial use, agricultural use, and military use. Fibrous clothing material is usually fabric finished or laminated so that it can function as waterproof, water-repellent, gas barrier, mist barrier, or dust barrier, according to its protective object. In the case of agricultural use, breathable waterproof and water-repellent clothing is usually desirable. Military gas barrier clothing is composed of fabric coated with nonpermeable material or laminated with activated carbon fiber fabric [47, 48]. The latter has advantage in the breathability over the former.

The basic principle for the protection of NBC is the same as that for chemical protection except for the additional protection of radioactive particles. The key point for radioactive protection is to prevent attaching of the particles on the fabric and sucking into the fabric [49].

5.5 The other kinds of protection and safety

As for other kinds of protection, we can mention (a) protection from injury of cutting/abrasion at hand works, (b) protection from metal splash and sputtering, (c) protection from cold environment, (d) protection from cold water immersion, (e) protection from hot environment, (f) sunlight protection, (g) electric static protection, (h) protection in the space, and (i) protection from electromagnetic wave.

As safety textile goods, there are (a) such goods related to fire accident such as hose, life chute, and rope ladder; (b) goods related to water accident such as life jacket and life raft; (c) goods related to impact damage such as helmet, safety shoes, and protectors as sporting goods; and (d) eye-catching mark attached to wears for safety.

In this part, some textile products related to some specific kinds of protection/safety are introduced.

5.5.1 Protection from cold environment

The primary measure of cold protection for clothing is thermal insulation by air layer within the fabric.

In cold foul weather, it must also act as a wind barrier. Insertion of reflective layer into the laminate for the radiation from human body is also effective. Exothermal heat of PCM contained in fibrous material, oxidizing heat of iron powder sustained in fibrous sheet, and electrical heating can also be available. But it must be noted that the effect of PCM is temporary.

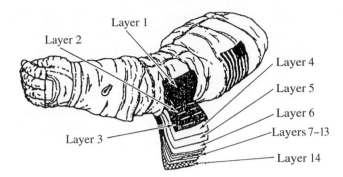

Figure 27. Composition of space suit [52].

5.5.2 Protection from hot environment

Protection from radiation heat is quite easy. But protection using textile products from hot atmosphere is quite difficult. Some trials using Peltier element have been carried out [50, 51]. PCM can be utilized for temporary cooling.

5.5.3 Space suit

Figure 27 shows a typical example of the space suit composition. In the figure, layer 1 is made of nylon woven fabric. Layer 2 is made of nylon/spandex fabric co-woven with thin tube in which water is circulating for thermal regulation. Layers 4 and 5 work as barrier layers for oxygen leakage whose materials are resin-coated fabrics. The roles of layers 6–13 are thermal insulation and protection against meteorite particle collision. Layer 14 is made of lamination by m-aramid fabric, porous polyfluoroethylene membrane, and p-aramid woven fabric with high yarn density, whose role is to protect from the collision and surface covering of the suit.

5.5.4 Protection from impact damage

In this category, there are two types of protective materials – flexible type and rigid type.

Regarding the flexible type, there has been almost no material that can effectively protect the human body from impacting damage by a projection having much larger curving radius than a bullet. But a system material of flexible type illustrated in Figure 28 has been recently developed. The central textile architecture of the material is capable of absorbing some impact energy. The material can easily stitch to such clothing as motorcycle wear.

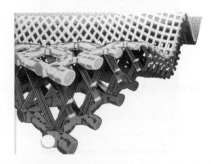

Figure 28. Flexible fibrous material protective from impact damage [53].

Composites can be useful as rigid-type material protective from impact damage. Helmet, the toe part of safety shoes, and some protectors for sports are representative examples based on rigid-type protection. There are two types of helmet by its material – fiber-reinforced plastic and thermoplastic (acrylonitrile/butadiene/styrene terpolymer resin etc.). The former can resist higher impact loading. As reinforcing fiber, glass fiber is usually used. But organic super fiber such as *p*-aramid is also used for reducing its weight. Safety shoe has a toe part that can resist high impact loading. Some examples of toe parts are made of fiber-reinforced plastics. The part of a commercial shoe, which is made of long fiber-reinforced thermoplastic, can sustain even the weight of standard passenger car [54].

6. Textile products for electric and information technologies

6.1 General scope of textile products in electric and information technologies

There are many kinds of textiles used for applications related to electric and information technologies. We can mention the following as the application examples: (a) print circuit, (b) battery and fuel cell, (c) insulation and cables, (d) electric conduction and shielding, (e) communication using optical fiber, (f) communication system parts utilizing fibrous materials, and (g) acoustic uses. In this section, items (a), (c), (d), (e), (f), and (g) are introduced. Battery and fuel cell are described in section 2.4.1.

6.2 Printed circuit

6.2.1 Printed circuit rigid board

The circuit is made by etching copper foil laminated on the board and/or by metal plating or by electric conductive paste printing on the board. The circuit is usually patterned by screen printing and/or photoresist method. Figure 29 is an example of printed circuit. In many cases, the circuit on one side of the board is connected to the circuit on another side through holes covered by metal plating. Recently, the board built up by circuits in multiple integrated layers has been developed.

The board is usually made of rigid paper or glass cloth impregnated by thermoset resin. In some cases, it is reinforced by woven fabric or wet-formed nonwoven of organic high-performance fiber.

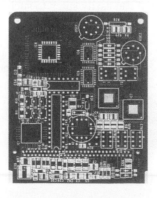

Figure 29. An example of printed circuit on the board [55].

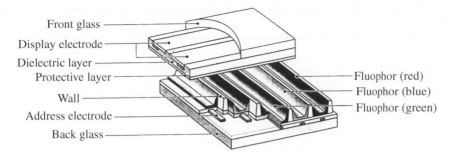

Front glass
Display electrode
Dielectric layer
Protective layer ———————————————— Fluophor (red)
Wall ———————————————— Fluophor (blue)
Address electrode ———————————————— Fluophor (green)
Back glass

Figure 30.   Structure of plasma display [56].

6.2.2 Screen for printing

Fine screen (mostly plain woven fabric made of monofilament) is one of the key materials of screen printing for electric devices. In many cases, the monofilament is made of sheath–core bicomponent fiber. Its typical examples of sheath/core are (a) nylon/PET, (b) toughened PET/PET, and (c) wholly aromatic polyester (see section 3.6 in the former book [1])/wholly aromatic polyester-reinforced flexible polymer. The screen made of (c) is the most feasible to precision screen print, because it has high resistance for mechanical deformation.

Screen printing using such materials is used for patterning of printed circuit, printing of fluorescent pigment for plasma display, as shown in Figure 30, printing dielectric paste for condenser, and printing of dam wall for liquid crystal of liquid crystal display [56].

6.3 Electric insulation and cables

6.3.1 Electric insulation materials

Electric insulation textiles are used for the insulation in several electric devices such as transformer and cables. They are in the form of nonwoven and woven fabrics impregnated in such liquids as varnish, mineral oil, unsaturated polyester resin, epoxy resin, silicone and polyimide resin, etc. As typical fiber material, cellulose, PP, nylon, PET, aramid, glass, etc., are used according to the required level of its thermal resistance. For the highest thermal resistance, glass-woven fabric impregnated by a liquid having high thermal resistance such as silicone, polyimide, polyimide-amide is used. In some cases, glass cloth impregnated in varnish is transformed into fabricated insulation parts [57].

6.3.2 Cables

Cables are used for several applications such as electric power transmission, telecommunication, and wiring for electric devices. According to the object of use, their structures are much varied. Relationships among the function of fibrous material, the kind of fiber material in cable, and its applied cable are summarized in Table 4. Figure 31 shows the cross-sectional structure of optical fiber cable used for main line.

6.4 Electric conductive materials and shielding materials

6.4.1 Electric conductive materials and their applications

There are several kinds of fibers having electric conductivity such as metal fibers, carbon fibers, fibers made of electric conductive polymer, metal plated organic fibers, organic

Table 4. Relationships among the function of fibrous material, used fiber material, and cable applied [58].

Function of fibrous material	Fiber material	Cable applied
Tension member	p-Aramid, glass	Optical cable
Tearing tool	p-Aramid, polyethylene terephthalate	Cable for wiring
Water absorption	Super absorptive fiber (see section 3.5.4)	Optical cable, copper cable
Thermal/fire resistance	Glass	Motor wiring cable, cable for fire fighting devices
Cushion, filling	Cotton, polyethylene terephthalate, nylon, rayon	Multicore cable
Reinforcement	Nylon, glass	Cable for aircraft wiring, cable for flectional motion

fibers coated with electric conductive material, organic fibers in which electric conductive particles are dispersed, and bicomponent fibers having electric conductive part in which electric conductive particles are dispersed.

Fibrous materials having electric conductivity are applied to such uses as antielectrostatic goods such as carpet, working wear, air filtration, conveyer belt, and clean room clothing, and to grounding within electric equipments, shielding of electromagnetic wave, circuit wire for electronic textiles, and the blush of electrostatic charging and discharging for copy machine and laser printer using electrophotograph.

6.4.2 Shielding of electromagnetic wave

The importance of electromagnetic shielding of electronic equipments has significantly increased. Fibrous materials can be effectively utilized for shielding. In this section, shielding technologies based on a fiber material developed by a Japanese company are introduced as examples of their utilization.

The electric conductive fiber is produced by electroless copper plating on PET fiber. Then it is further plated by a thin layer of nickel alloy for chemical resistance and for

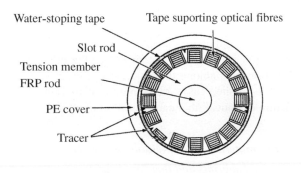

Figure 31. Optical fiber cable of slot type used for main line [59]; FRP indicates fiber-reinforced plastic; PE, polyethylene.

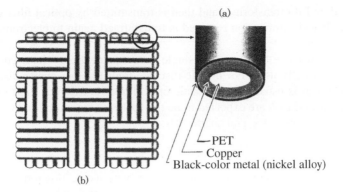

(a)

—PET
— Copper
Black-color metal (nickel alloy)

(b)

Figure 32.   An example of electric conductive fiber and its woven fabric [60]; PET indicates polyethylene terephthalate.

making its color black, as shown in Figure 32(a). The fabrics (Figure 32(b)) made of the fiber have the shielding efficiency of about 80 dB. Some of them are transformed to adhesive tape, which can be easily applied to cover plate, and the box to be shielded. The sheet composed of its knitted fabric, electric conductive urethane foam, and its nonwoven are used for electric grounding by inserting between circuit boards. The screen made of the fine monofilament fiber is laminated to plasma display panel to cut the excessive wave emitted from plasma display, as shown in Figure 33.

6.5 Textile products related to communications

6.5.1 Optical fiber for telecommunication

The telecommunication by optical fiber (see section 6.2 in the former article [1]) has great advantages in the capacity of information transmittance, and in the security for outer disturbance and information diffusion over those of the other media. Electronic information

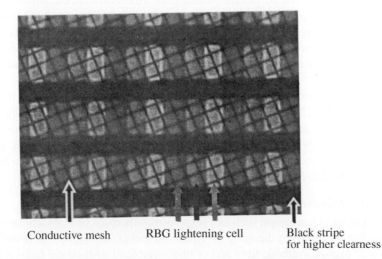

Conductive mesh   RBG lightening cell   Black stripe
                                                  for higher clearness

Figure 33.   Electromagnetic shielding of plasma display by mesh cloth [61]; RBG indicates red, blue, and green.

is transformed by E/O transformer and then is transmitted by optical fiber cable through some relays. Optical information received is transformed again to electronic information by O/E transformer.

For the long-distance transmission, single-mode quartz fiber whose core diameter is about 10 μm is usually used. For middle-distance transmission, quartz fiber whose core diameter is 200 μm is usually used. Plastic optical fiber is used for short-distance transmission. Its main end uses are expected to be the LANs of automobiles, buildings such as offices, hospitals, universities, and apartments.

6.5.2 Optical fibers for other uses

We can mention (a) image guide such as fiber scope, pattern detecting rod for facsimile, and scanner; (b) lighting and illumination; and (c) sensing as the other uses of optical fibers. Plastic optical fiber is used for most of them. In the case of (a), fiber is usually used in the form of a fiber bundle. The sensing using optical fiber is generally useful for watching an unusual phenomenon in such cases as structural systems, land, and factory products.

6.5.3 Materials for satellite

Main structures are usually made of honeycomb sandwich in which carbon woven fabric-reinforced polymer composites are used. Solar panel is usually supported by thin board made of carbon fiber-reinforced polymer composite.

Triaxial woven fabric-reinforced composites are especially feasible to reflector antenna, because they are the thinnest among quasi-isotropic sheets. Recently, inflatable antenna and other structures made of triaxial woven fabric composites, which are expanded by spring back from folded form has been developed. In these uses, carbon fibers of high modulus type are especially used to realize higher stiffness of composites.

6.6 Acoustic uses

6.6.1 Speaker diaphragm

Material of speaker diaphragm is required to have high tensile modulus, high internal friction loss, and low specific gravity for getting good vibration characteristics. Therefore, a fiber such as carbon fiber, organic fiber having high modulus, and glass fiber is used for its reinforcement. The examples of high-quality diaphragm are (a) composites sheet composed of carbon fiber and thermoset resin, (b) glass fiber-reinforced thermoplastic board made by injection molding, and (c) paper blended with such a high-performance fiber as those mentioned above [62].

Screen integrated with speaker that is made of PEN (see section 11.4.1 in the former article [1]) woven fabric has been developed. The screen can be an image display medium and speaker diaphragm at the same time, as shown in Figure 34. The high modulus and reasonable price of PEN fiber make the material suitable for this application. Its sonic features are (a) simple system, (b) clear-cut sound quality by its high-vibration damping, and (c) no sound directivity with higher sound pressure [63].

6.6.2 Musical instrument parts

Felt used for a musical instrument such as a piano is usually made of wool whose functions are (a) providing cushion, (b) damping of vibration, and (c) absorption of sound. But in some electronic pianos, synthetic fiber and/or fiber combined with elastic foam is used for its felt part [64]. For the stringed instrument, several kinds of materials such as silk yarn, gut, synthetic fiber, and steel wire are used [65].

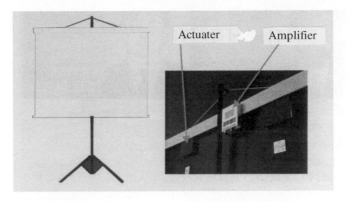

Figure 34. Screen speaker [63].

7. Textiles for construction and civil engineering

7.1 General scope of textiles in construction and civil engineering

There are several kinds of textiles used for construction and civil engineering, such as (a) fiber reinforcement of concrete and polymer composites for construction uses; (b) fiber for antiseismatic reinforcement; (c) fabric membrane for building; (d) reinforcement by flexible tube for gas pipe; (e) reinforcement of concrete pipe for water service; (f) geotextiles for the reinforcement of ground, slope, and road; (g) geotextile for such uses as drainage, protection of water corrosion, separation, and fabric form; and (h) fibrous materials for heat insulation, sound control, moisture control, and waterproof of construction uses. In this chapter, these textiles are introduced by the following three sections: (1) fiber reinforcement for construction/civil engineering uses containing (a) and (b), (2) geotextiles containing (f) and (g), and (3) miscellaneous construction/civil engineering materials containing (c), (d), (e), and (h).

7.2 Concrete/cement reinforcement

7.2.1 Short-fiber reinforcement of concrete by blending

Short fibers are used for reinforcement of concrete in dispersed blending, whose volume fraction is less than a few percent. Its major fiber materials are polyvinyl alcohol (PVA) (see section 11.1 in the former article [1]). But in some cases, such fiber materials as alkali-resistant glass, PP, acrylonitrile, and carbon are used. The fiber-reinforced concrete is applied to construction through such a method as the use of prefabricated board, filling to form work, or spraying. The increases in tensile strength and bending strength by the reinforcement are not so significant. But the reinforced concrete has 10 to 20 times higher breaking energy (toughness) than the unreinforced cement [66].

Highly toughened concrete has been developed by using high-tenacity PVA fiber of high thickness, which is produced by solvent gel spinning method. It can be nailed and has toughness almost equivalent to aluminum board, as shown in Figure 35(a). The mechanism of such a high toughness is related to the growth of fine cracks, as shown in Figure 35(b), which can be explained by the theory named 'ECC' by V.C. Li [68]. It has been applied to several constructions such as the floorboard of a large bridge, the connecting beam of a large building, and the wall surface of a tunnel.

(a)

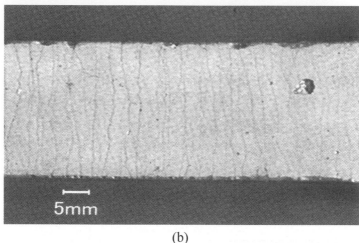

5mm

(b)

Figure 35. The deformation behaviors of concrete board reinforced by high tenacity PVA short fiber [67]: (a) bending test of the concrete board and (b) microcracks of the concrete board grown by tensile load.

7.2.2 Reinforcement aligned fiber structures

There are three kinds of major reinforcements using aligned fibers: alternative to steel or prestressed steel reinforcement of concrete, alternatives to steel beam, and reinforcement of wall panel.

7.2.2.1. Alternatives to steel or prestressed steel reinforcement of concrete. As an alternative material to steel for the reinforcement of concrete, there are continuous fiber-reinforced polymer or cement composites for which fiber having high tensile modulus, such as carbon fiber and *p*-aramid fiber, is used. Figure 36 shows examples of the reinforcement materials. Most rodlike materials have some rugged surface, as shown in the figure. Some of them take such forms as three-dimensional lattice and hoop. In some cases, they are used as

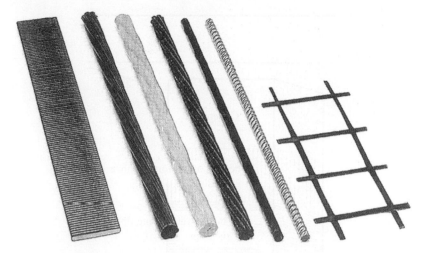

Figure 36. Fiber-reinforced alternatives of steel reinforcement of concrete [69].

prestressed reinforcement in order to effectively utilize their resistance to tensile loading within concrete beam.

They have the following features compared with steel reinforcement: light weight, no corrosion, and no electromagnetic wave interference (except the case of carbon fiber reinforcement).

7.2.2.2. Alternatives to steel beam. Fiber-reinforced polymer composite beam is used as an alternative to steel beams in some cases. It is used for roof structure usually by a pipe truss, because of its lightness. It is also used for bridge beam and its floorboard.

7.2.2.3. Reinforcement of concrete board. There are concrete boards reinforced by woven fabrics. They are used for such applications as internal wall, external wall, floor, ceiling, and formwork board. Their thickness can be less than that of the nonreinforced board.

7.2.3 External reinforcement

It is used for the reinforcement of existing constructions. Reinforcement for preservation of historical construction, aseismatic reinforcement, and refreshment of old constructions are its typical applications. The reinforcement is conducted by winding of the tape or strand of high-mechanical-performance fiber, such as carbon fiber impregnated by resin, on pillars. In some cases, the tape is used by winding and laminating to pillars, as shown in Figure 37. The board of polymer or cement composite reinforced by such a fiber is also used by laminating or mechanically fixing to pillars and walls. Figure 38 is an example of pillar reinforcement using short-fiber-reinforced concrete board and mechanical fixing by the winding of steel wire.

7.3 Geotextiles

In this section, geotextiles are defined as materials for civil engineering in which fibrous materials are used at least as one of its key materials.

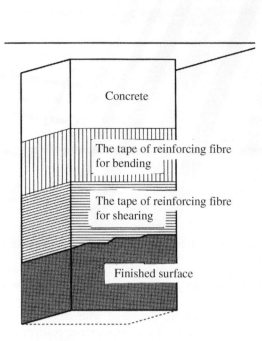

Figure 37. Pillar reinforced by making use of tape [70].

7.3.1 Reinforcement of filled slope, soft ground, and mud

Figure 39 shows the reinforcement of slope using geotextile sheet. The weight of earth tends to cause a slippage at the shoulder along circular face. The key factors of the sheet to prevent such a slippage are its tensile loading capability and the friction force between earth and the sheet. Figure 40 is an example of reinforcement structure for sharp filled earth using geotextile sheet, gabion (this is also a kind of geotextile), and concrete. In

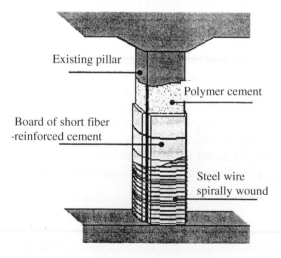

Figure 38. Pillar reinforced by making use of board and wire [71].

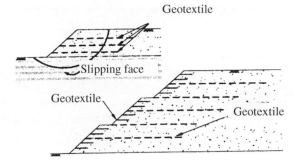

Figure 39. Reinforcement of slope by geotextiles [71].

several earthquakes that recently happened in Japan, it was proved that reinforcements using geotextiles are far better than other kinds of reinforcements. Figure 41 illustrates a typical reinforcement of soft ground by geotextile sheet. By laying the sheet, bulldozer can operate on even muddy ground.

Spun-laying nonwoven and geo-grid are usually used as a geotextile sheet for this kind of reinforcement. The geo-grid belonging to geotextiles is usually fabric of mesh structure.

There are still other kinds of reinforcements using geotextiles. The following are some of their examples: (a) short-fiber reinforcement of earth by blending; (b) blending short-fiber material and absorbent particles with mud; (c) filling earth into textile bag; (d) reinforcement of slope by bump jacket anchor composed of textile circular rugged tube, steel rod, and grout intruded into the tube, as illustrated in Figure 42; and (e) reinforcement of slope and muddy ground by specific sheet combined with jacket filled with grout. The features of the specific sheet are that it can be very rigid after the filling of grout and that its setting work is fairly simple.

7.3.2 Materials for drainage, separation, filtration, protection, and water barrier

7.3.2.1. Drainage. There are several kinds of geotextile material for drainage, as shown in Figure 43. Some examples of their materials are as follows: spun-laid nonwoven and gravels, in Figure 43(a), tube made of monofilament in (b), monofilament nonwoven board covered by spun-laid nonwoven in (c), and plastic-corrugated board laminated with nonwoven in (d).

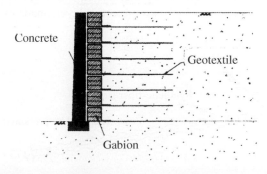

Figure 40. Sharp filled earth using geotextiles and concrete [71].

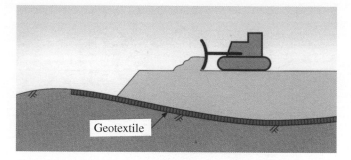

Figure 41.   Reinforcement of soft ground by geotextile sheet [72].

There are vertical drainage and horizontal drainage in the earth. There are also drainages on the slope and on the wall. Figure 44 is an example of vertical drainage system making use of vacuuming.

7.3.2.2. Separation. The function of separation is to separate two different layers such as (a) two kinds of earth layers, (b) earth layer and small stone layer, or (c) water, silt, etc. In many cases, a geotextile has separation function and the other function such as reinforcement, protection, and filtration at the same time. In the example of Figure 41, the geotextiles sheet has the roles of both reinforcement and separation. But we can mention silt protection membrane and pumping protection as typical application examples related to separation function. Figure 45 shows an example of railway cross section using four kinds of geotextile sheets: for pumping protection, for ballast net, for reinforcement, and for grass control.

7.3.2.3. Filtration. In the case of drainage, water flows along the sheet surface. But in the case of filtration, water flows out across the sheet through which earth particles are filtrated. A typical example of geotextile related to filtration function is the prevention of the sand flowing out by suction at shore protecting work, as illustrated in Figure 46. Nonwoven or fiber-reinforced nonwoven is usually used for these kinds of sheets.

7.3.2.4. Protection. Geotextiles belonging to this category are mainly used for protecting water-barrier membranes of such uses as pond, tunnel, and landfill place. Figure 47 is an example of use for landfilling in which two protective layers made of thick nonwoven are

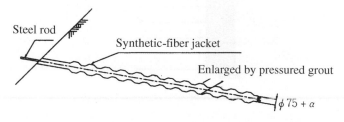

Figure 42.   Reinforcement by expandable anchor bolt using textile jacket [73].

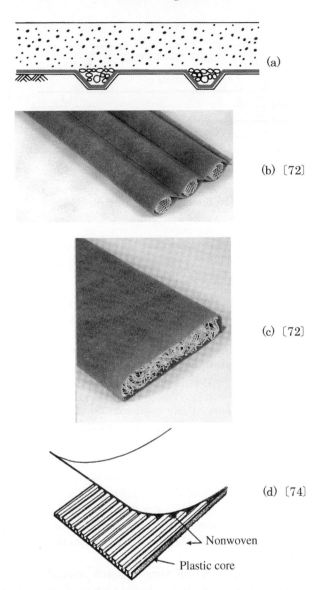

(a)

(b) 〔72〕

(c) 〔72〕

(d) 〔74〕

Nonwoven

Plastic core

Figure 43.   Several kinds of drain/drain materials.

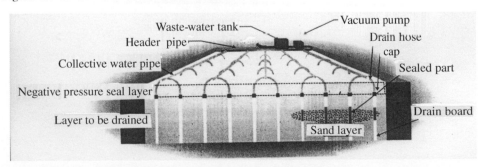

Figure 44.   Vertical drainage system using drain board [75].

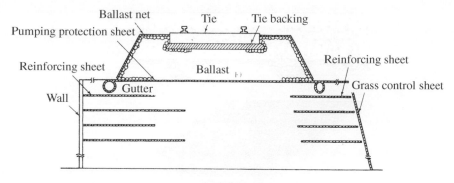

Figure 45.   An example of railway cross section using geotextiles [76].

used. Protections from mechanical damage and light degradation are the main functions of the surface layer.

7.3.2.5. Water barrier. Some water-barrier sheets are made of fiber-reinforced rubber. They are often used in double layers with protective fibrous sheets for landfill place, as shown in Figure 47. Another structural example for water barrier is composed of two protective nonwoven sheets, two water-barrier rubber sheets, and a porous drainage layer in which sensors to watch water leakage are installed.

7.3.3 Materials for some other kinds of geotextile uses

There are geotextiles for the other kinds of uses such as fabric formworks, sheet for greening to protect riverbanks from erosion caused by water flow, artificial turf, bag net for containing stones, and flexible tubular dam. Figure 48 shows some types of fabric formworks in which concrete is filled. In the Figure, (a) is water sink type, (b) is grid type for greening, and (c) is water-barrier type. Figure 49 is the sheet for greening to protect riverbank. The part of grass near its root is guarded by the sheet from water flow, and thus, the riverbank covered by the sheet can be protected from erosion. Figure 50 is a dam system using flexible tube that is made of fabric-reinforced rubber. The height of the dam can be automatically controlled by its internal pressure.

Recently, a new system using optical fiber to watch unusual strain growth for civil engineering construction has been developed. In the system, if optical fiber that is settled

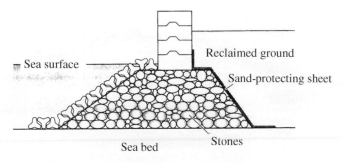

Figure 46.   An example of protection for sand flow out at sea shore work [77].

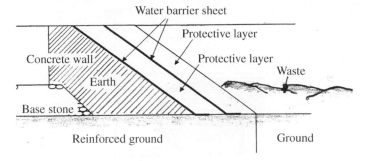

Figure 47.   An example of wasteland filling place using geotextiles [78].

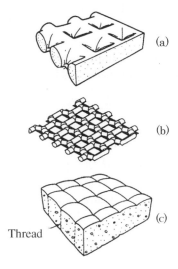

Figure 48.   Various fabric form works [79].

Figure 49.   Sheet for greening to protect riverbank from erosion caused by water flow [80].

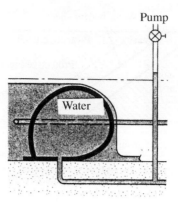

Figure 50.   Dam made of flexible tube [81].

along the construction is strained, the light frequency of Brillouin backscattering is shifted. Thus, such events in the range of 10 km as the crack of tunnel and the microslip of earth at slope can be monitored.

7.4 Miscellaneous construction/pipeline materials

7.4.1 Materials for sound control and waterproof in construction

7.4.1.1. Sound control. There are three kinds of materials for sound control: (a) sound absorption board, (b) sound insulation board, and (c) antiacoustic vibration sheet. Glass wool is the main material for sound absorption board, which usually has dressed face made of such material as glass cloth and polyvinyl chloride sheet. It is used for the ceiling and the walls of offices, hotels, schools, hospitals, etc.

7.4.1.2. Waterproof in construction. The waterproof materials for roof of building can be classified into asphalt, resin-coated sheet, and thick paint coating. But the most popular method is asphalt roofing. In this method, the nonwoven made of synthetic fiber is used for reinforcing and stabilizing the asphalt layer.

Recently, the paper sheet that can be breathable and nailed with keeping high water tightness by using super water absorbable fiber has been developed for waterproof roofing of woody house [82]. It is used by laying under-roof tiles.

7.4.2 Material for membrane construction

Representative material for big membrane construction is usually composed of glass fiber plain woven fabric coated with polytetrafluor oethylene (PTFE) resin as illustrated in Figure 51. On the other hand, the material for small membrane construction is usually PET fabric coated with polyvinyl chloride. Membrane construction can be classified into pneumatic structure and tensioned structure using rigid bones. Typical examples of the former are air domes. An example of the latter is shown in the bottom illustration of Figure 51. The need of membrane construction having open structure for UV protection has increased by the matter of health.

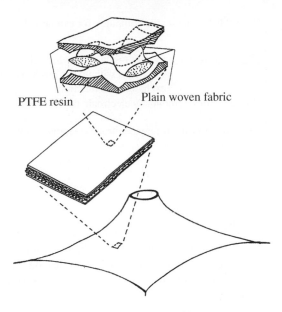

PTFE resin   Plain woven fabric

Figure 51. Material structure of membrane construction [83].

7.4.3 Materials for pipelines

There are several kinds of pipelines for such uses as water supply, drainage, irrigation, gas supply, electric power supply, and telecommunication. Hose made of circular woven fabric coated with elastomer resin is used for the renewal of these lines made of steel pipe. The hose is inserted into old steel pipe to be renewed by inverting using pressured air as illustrated in Figure 52. In some cases for such uses as drain, gas supply, the hose impregnated with B-stage epoxy is used. In these cases, after its insertion, curing is conducted by steaming and then fiber-reinforced polymer composite rigid pipe fixed to the internal surface of old steel pipe is formed.

For new settled pipelines, plastic pipe has replaced the steel pipe because of its low maintenance cost. But it cannot be used for the line that should bear high inner pressure and/or high apex pressure. The pipe composed of plastic internal surface layer, fiber-reinforced concrete layer, and concrete outer layer has been developed for such pressured pipelines [85].

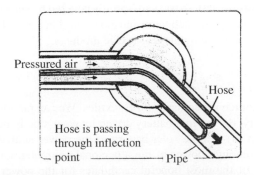

Pressured air

Hose

Hose is passing
through inflection
point   Pipe

Figure 52. Insertion of hose into pipe by inversing using pressured air [84].

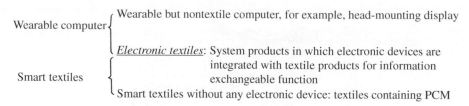

Figure 53. Definition of E-textile and its differences from wearable computer and smart textiles; PCM indicates phase-change material.

8. Electronic textiles

8.1 General scope of electronic textiles

It must be reasonable to define "electronic textiles (E-textiles)" as system products in which electronic devices are integrated with textile products for information exchangeable functions. We can mention "wearable computer" and "smart textiles" as concepts similar to E-textiles. But it must be noted that there are some differences among them, as schematically shown in Figure 53. E-Textiles are also a kind of computer network system product close to the human body along ubiquitous environment.

In the former part of this chapter, materials, parts, and circuit systems for E-textiles are introduced. In the latter part, several application systems of E-textiles are overviewed.

8.2 Materials, parts, and circuit systems for E-textiles

8.2.1 Circuit materials and circuit systems for E-textiles

The necessary conditions to ensure that the circuit is feasible to E-textiles are (a) the circuit material must be flexible enough, (b) it should have enough mechanical toughness, and (c) it should have enough washability. It is also desirable that it is stretchable and can be colored. We can mention the following wiring methods for the circuit: (a) weaving or knitting of electric (or photo) conductive yarn into the cloth, (b) embroidering or stitching of such a yarn on the cloth, (c) printing of conductive material on the cloth, and (d) laminating of conductive tape on the cloth.

There are electric conductive yarns, which are mixed with ultra-fine metal fiber and organic electric conductive fiber (see section 7 in the former article [1]). Recently, electric conductive fiber made of conventional material such as PET and PVA, whose conductivity is at the level of 1 Ω/cm has been developed [86,87]. Electric conductive and stretchable yarn whose core consists of elastomeric yarn spirally covered by such an electric conductive fiber have also been developed. Such yarns must be useful because they can be stretchable and dyeable and have enough toughness as textile products.

8.2.2 Power sources for E-textiles

At least some power source is necessary for E-textiles. We can mention such candidates as (a) conventional battery, (b) thermal battery driven by wear's body heat, (c) battery driven by wear's motion, and (d) flexible solar battery as the power source. It is said that solar battery does not have enough power to run a personal computer. Organic radical battery [89] seems to be one of the most hopeful candidates for the power source because it is extremely thin and flexible.

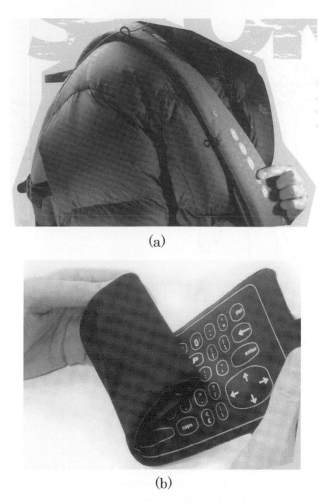

(a)

(b)

Figure 54. Examples of fabric parts for switching and controlling [91]: (a) a controller part for portable audio player and (b) a flexible keyboard made of fabric.

8.2.3 Transmitting and receiver devices and interfacing technologies for E-textiles

Microslit antenna for satellite communication and transmitter using fabric substrate have been developed. Recently, the technologies of communication through the human body have been developed. Its features are no disturbance by external interference and easy communication by only human touching action. One of the technologies utilizes the changes of electric field at the surface of the human body by a specific photonic electric field sensor [89]. This kind of technology seems to be one of the hopeful communicating and interfacing tools for E-textiles.

There can be such input tools for E-textiles as touch panel, keyboard, and audio instruction. Figures 54(a) and (b) are examples of input tools that have been commercialized.

Visual output tools are ranged from an indicating lamp to image display panel for visual. Audio output is more desirable for some useful purposes. For wearable computer, head-mount display is conventionally used for image display. Figure 55 illustrates a sleeve display integrated into a working wear and a watch-type controller.

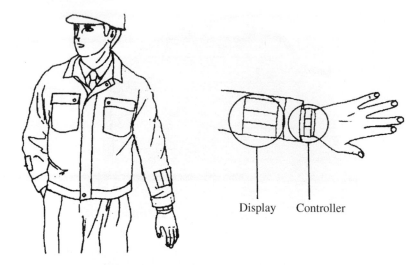

Figure 55. A sleeve display integrated into a working wear [91].

Recently, a robot suit system utilizing an innovative interfacing technology has been developed by Sankai [92]. In the technology, the signal of very small amount of bio-ion current from his brain to move the muscle is detected by sensors attached to his muscle. On the basis of the signal, a small-sized computer settled on his back instantly orders to start the movement of the suit part related to his muscle. The fact must be noted that the wearer and the suit are intimately integrated in this system.

8.2.4 Sensing technologies for E-textiles

Ghosh gave a perspective article on E-textiles in which several kinds of sensing technologies for such objectives as motion, physiological condition, damage of human body, and chemical protection are introduced [93].

Several technologies for location sensing have been tried [94, 95], but these have been unsuccessful in practical sensing for the wandering location of dementia in elderly people.

Sensing of human motions is an important item to be studied for E-textiles related to rehabilitation, sport training, and assistance in human works. In most cases, it is based on changes in electric resistance, which is caused by the deformation of elastomeric substrate attached on the human body surface.

Sensing of human physiological condition and psychological condition through E-textiles must be useful for both health monitoring and medical treatments. Several studies have been carried out for E-textiles on sensing changes in such physiological parameters as (a) electric resistance and temperature of human skin, (b) blood flow rate, (c) heart rate, (d) electrocardiogram, and (e) motions [93, 96–98].

8.2.5 Polymer photonic fiber for E-textiles

Polymer photonic fibers and fabrics that have functions of generating, transmitting, modulating, and detecting photons have been developed. They can be used for sensors, wave guide, and pattern display of E-textiles. Photosensitive polymeric optical fiber was irradiated by UV laser, and then Bragg grating structures were fabricated in the fibers [99].

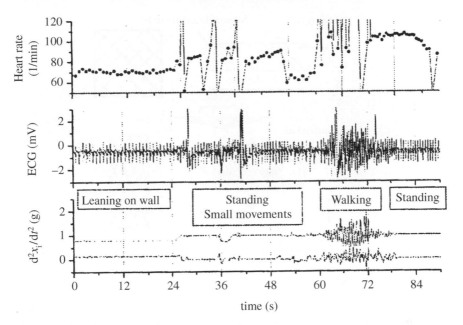

Figure 56. An example of simultaneous observation results on heart rate, electrocardiogram (ECG) and motional acceleration [101].

8.3 Application systems of E-textiles

8.3.1 Audio entertainment

Clothing integrated with audio player system and portable phone, such as that shown in Figure 54, has been commercialized.

8.3.2 Medical monitoring of physiological conditions

Research on medical monitoring systems must have been most intensively carried out among the application researches of E-textiles. Sensors for monitoring physiological condition items necessary for medical judgment are attached to underwear. Information collected by these sensings is transmitted to data processing unit. Monitoring these data is continuously or intermittently conducted.

Portable phone network systems can be utilized for the transmission [100]. Figure 56 is an example of simultaneous observation data of heart rate, electrocardiogram, and motional acceleration rate. Research on monitoring system for rehabilitation motions of patients has also been carried out [102].

Some monitoring systems for practical trial have been conducted. The usability of a system for long-distance monitoring of heart physiological conditions of patients who are under rehabilitative process was tested. In the system, the data are wirelessly transmitted to a central monitoring computer consisted of Web module, database module and diagnosis module. If some problem happens, its signal is transmitted to the light-emitting diode (LED) lamp and alarm of the patient for alarm [103].

8.3.3 Assistance of human power/motions

Several kinds of systems for enhancing human power needed in workings have been developed, which are generally called as wearable robot. For these systems, several kinds

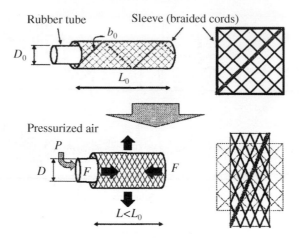

Figure 57.   The structure and working principle on McKibben type man-made muscle [104].

of the motive-power sources such as man-made muscles and motor have been utilized. Figure 57 shows the working mechanism of a typical man-made muscle, which is called a McKibben type driven by pressurized air. Another example of a wearable robot driven by a motor, which is previously introduced in section 8.2.3, is shown in Figure 58. The procedure for lifting work using the muscle system includes (a) thinking to lift up the object by hand, (b) electric signal is transmitted from brain to human muscle, (c) sensors on human arm surface detect biovoltage, (d) computer forecasts to do lifting up motion with the biovoltage, (e) computer orders to operate the motor of power unit, and (f) just before the human lifting up motion, the motor starts to assist the human lifting up motion. It is expected that these systems are useful for care working, rehabilitation, military use, and rescue working. But strictly speaking, this robot suit cannot be classified as an E-textile because any textile is not used as a key material in the suit.

Recently, a robot suit system assisting the rehabilitation of patients who have half their body paralyzed has been developed. Man-made rubber muscle driven by pressured air is integrated into the arm part of the suit. The muscle is actuated to move the paralyzed arm in the same manner as a normal arm by detecting the motion, as shown in Figure 59.

8.3.4 Intelligent assistance of workings

We can find some reports on the systems of intelligent assistance of workings using wearable computer. Averett introduced several examples of their practical uses for such workings as order picking, maintenance of aircraft, medical operation, and construction [107]. They belong to wearable computers but cannot be classified as E-textiles except oral communication system integrated to working wear protective against very cold environments.

There have still been some examples of research on working wear system through which wearers can communicate with each other and the commander [108,109]. Such systems seem to be useful for insuring the worker's safety in dangerous work environments and cooperative workings with complicated and harmonized operations in each separate position.

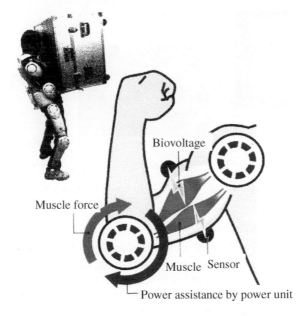

Biovoltage

Muscle force

Muscle   Sensor

Power assistance by power unit

Figure 58.   Wearable robot for assisting human power in working [105].

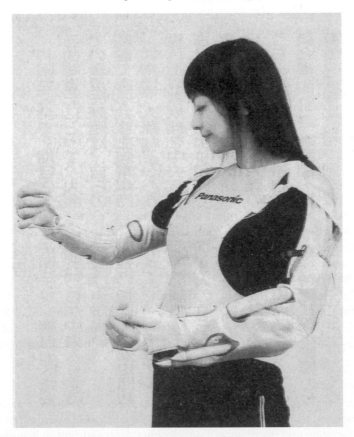

Figure 59.   Robot suit assisting rehabilitation of patients who have half their body paralyzed [106].

Figure 60.   Examples of decorative uses based on light-emitting pattern by an E-textile system [110].

8.3.5   Fashion, decoration, and advertisement

By docking some units for output by such forms as signal light, display, and sound with textiles, we can positively make use of them for fashion, decoration, entertainment, advertisement, and safety visibility. This kind of E-textile has the following significant potentials by their nature – to be flexible, dynamic, modular, and interactive.

A system that has LED array display unit driven by designed plan input into personal computer using specified software has been developed. Figure 60 shows some examples of its decorative uses. Several trials of application to fashion [111], decoration, advertisement, and the on-stage action have also been tried.

8.3.6   Assisting sports and actions for health

A system of training wear for running by which heart rate, speed, and distance can be monitored by using the sensors snapped at its heart position and shoes has been commercialized. Data are sent to a wrist-mounted computer, which can display and record in real time [112]. A wearable health monitoring system in which a heart rate sensor is attached to a sports shirt has also been commercialized [113].

8.3.7   Positive control of microclimate temperature

In this section, "positive control" means that the system utilizes at least some electrical devices for control. For heating, we can conventionally utilize electric resistance heating. There are many textile systems using this type of heating for thermoregulation of such products as carpet, blanket, and gloves. On the other hand, Peltier cell can work for both cooling and heating, as schematically shown in Figure 61. But in the system using Peltier cell for cooling, the problem is how to effectively and practically liberate heat from the external surface of the cloth. In a system, heat sink made of stainless steel filaments and small fans is adopted for the liberation [50]. Such a cooling system is useful for work clothes in very-hot-air environment and for clothes of congenital anhidrosis patients.

8.3.8   Intelligent tag

A textile intelligent tag has been developed. It is washable and has transmission and receiving functions, as well as a computer chip. By the tag, the recording of the owner's name, manufacturing and distribution data, cleaning data, and care indication is possible [114].

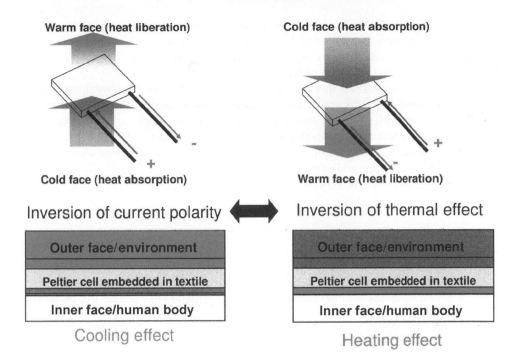

Figure 61. Principle of thermoregulation using Peltier cell embedded in textiles [50].

8.3.9 Intelligent interiors

A system of intelligent carpet has been developed, which can be used for a guide of the way, a control panel for lights and air conditioning, fire alarm, indication of room, and setting robot-moving route [115]. Such technology can be easily extended to other kinds of interiors.

8.4 Future perspectives

Presently (in 2008) E-textiles are still at the developmental stage as a whole, but it can be understood that there is a big potential for E-textiles if existing technological elements are effectively utilized for the system that is properly designed to answer social potential needs. Key points for their success must be (a) patient and successive efforts of developmental activities, (b) finding negotiable solutions for customers between their cost and performance, and (c) realizing their easier handling for customers.

9. Concluding remarks

While writing this article, the author took special care of the following: (a) systematic introduction by setting the general scope for each chapter or some sections, (b) systematic constitution of contents, (c) orientation to only applied products, (d) introduction of advanced technologies, but not too advanced technologies from a practical point of view, (e) orientation to technologies, but to be based on scientific knowledge, and (f) the use of many figures and tables for easier understanding for readers. One section in this article is dedicated to E-textiles, which are now rapidly progressing. Hence, the author is afraid that this section might be outdated in a few years.

As described in the beginning of this article, it is published successive to the former article *Fiber Materials for Advanced Technical Textiles* [1] in the advanced textile series. Hence, the author would like to suggest that readers who want to have the complete knowledge of advanced technical textiles, read both articles.

The author is also afraid that the readers may counter some inconvenience for the literatures referred to in this article because many of them are written only in Japanese. But this fact also brings one advantage here, because it can supply a lot of information on advanced technical textiles developed in Japan to the readers who do not have easy access to Japanese literature. In order to compensate the inconvenience to some extent, a few books are listed at the end of the references.

Finally, the author will be very pleased if readers can read this article with an intellectual satisfaction.

References

[1]   T. Matsuo, Text. Prog. 40 (2008) p. 87.
[2]   C. Byrne, *Handbook of Technical Textiles*, The Textile Institute, 2000, p. 11.
[3]   T. Senda, *Handbook of Fibres (Sen-I Binran)*, The Society of Fiber Science and Technology, Japan, Maruzen Co., 2004, p. 893.
[4]   J. Kamo, Seikei-Kakou 17 (2005) p. 300.
[5]   TOYOBO NOW, No 7&8, Toyobo Co., Osaka, Japan, 2007, p. 15.
[6]   T. Ishikawa, *A new fabrication method for functional fiber having nano-sized surface layer and its application system*, Abstractive lecture notes of spring seminar by The Textile Machinery Society of Japan, 2004, p. 67.
[7]   T. Kato, *Handbook of Fibres (Sen-I Binran)*, The Society of Fiber Science and Technology, Japan, Maruzen Co., 2004, p. 895.
[8]   A. Wada, *Handbook of Fibrous Materials for Industrial Use*, The Society of Fiber Science and Technology, Japan, 1994, p. 286.
[9]   Y. Iizuka, *Handbook of Fibrous Materials for Industrial Use*, The Society of Fiber Science and Technology, Japan, 1994, p. 312.
[10] E. Wadahara et al., Seikei-Kakou 19 (2007) p. 745.
[11] F. Ko, *Textile Structural Composites, Three-Dimensional Fabrics for Composites*, Elsevier, Amsterdam, 1989, p. 142.
[12] Y. Izuka, *Automobile and Textiles, Application of Composites to Car Parts*, Sen-I Sha Co., Osaka, Japan, 2004, p. 181.
[13] Y. Izuka, *Fundamental and Practical Aspect of Advanced Composites*, Vol. 40, Kakou-gijutsu, Sen-I Sha Co., 2005, p. 631.
[14] O. Aoki, *Automobile and Textiles, Present Aspect and Future Prospect on Lightening of Automobile*, Sen-I Sha Co., 2004, p. 28.
[15] O. Aoki, *Automobile and Textiles, Present Aspect and Future Prospect on Lightening of Automobile*, Sen-I Sha Co., 2004, p. 31.
[16] Y. Izuka, *Automobile and Textiles, Application of Composites to Automobile*, Sen-I Sha Co., 2004, p. 183.
[17] H. Horiuchi, Sen'i Gakkaishi, 59 (2003) p. 207.
[18] Y. Yamasaki, *Challenge for Innovations in Textile Industry, Frontier of 3R in Textile Products*, Sen-I Sha Co., 2002, p. 175.
[19] T. Yoshihara, *Recycle of nylon 66 fabrics to engineering plastics*, Lecture at Textile Recycle Research Committee of The Textile Machinery Society of Japan, 2002.
[20] M. Tsue, *Automobile and Textiles, Noise Control Technology Using Shredder Dust*, Sen-I Sha Co., 2004, p. 266.
[21] The pamphlet of Cyberlon™, Asahi Kasei Co., Osaka, 2006.
[22] A. Yonenaga, private communication, 2006.
[23] K. Kawasaki, *Automobile and Textiles, Tyre and Tyre-Cord*, Sen-I Sha Co., 2004, p. 266.
[24] K. Kawasaki, *Handbook of Fibres (Sen-I Binran)*, The Society of Fiber Science and Technology, Japan, Maruzen Co., 2004 p. 660.

[25] T. Kinoshita, *Automobile and Textiles, Driving Belts*, Sen-I Sha Co., 2004, p. 167.
[26] A. Yonenaga, *Automobile and Textiles, Textiles for Automobile and Their Trends in Europe*, Sen-I Sha Co., 2004, p. 278.
[27] H. Endo, *Automobile and Textiles, Floor Materials*, Sen-I Sha Co., 2004, p. 133.
[28] A. Hinata, *Automobile and Textiles, Nonwovens for Automobile*, Sen-I Sha Co., 2004, p. 74.
[29] Y. Hama, *Automobile and Textiles, Engine Filters*, Sen-I Sha Co., 2004, p. 203.
[30] K. Maruki, *Automobile and Textiles, Oil Filters*, Sen-I Sha Co., 2004, p. 208.
[31] S. Minemura, *Automobile and Textiles, Air Purification of Cabin*, Sen-I Sha Co., 2004, p. 212.
[32] M. Sibuya, *Automobile and Textiles, Filter for Automotive Exhaust Gas*, Sen-I Sha Co., 2004, p. 217.
[33] T. Matsuo, Text. Prog. 40 (2008) p. 97.
[34] A. Hinata, *Automobile and Textiles, Nonwovens for Automobile*, Sen-I Sha Co., 2004, p. 82.
[35] M. Yokogi, Sen'i Gakkaishi 55 (1999) p. 340.
[36] T. Kawaguchi, Sen'i Gakkaishi 58 (2002) p. 320.
[37] J. Kamo, Sen'i Gakkaishi 60 (2004) p. 482.
[38] P.A. Smith, *Handbook of Technical Textiles*, The Textile Institute, Woodhead Publishing Ltd., 2000, p. 130.
[39] H. Shoji, Sen'i Gakkaishi, 61 (2005) p. 217.
[40] Unitica Co. Ltd., WEB J. 44 (2002) p. 22.
[41] Y. Qin, Text. Mag. 1 (2005), p. 12.
[42] Bus. Ind. News (March 2007) p. 6.
[43] *Technology Trends of Functional Fibers*, Toray Research Center, 2005, p. 304.
[44] Y. Nomura, *Future Textiles, Bullet Proof and Stab Proof*, Sen-I Sha Co., 2006, p. 329.
[45] Y. Yamamoto, *Handbook of Fibres (Sen-I Binran)*, The Society of Fiber Science and Technology, Japan, Maruzen Co., 2004, p. 544.
[46] M. Fujiyama, *Future Textiles, Fire Fighting Suit*, Sen-I Sha Co., 2006, p. 319.
[47] S. Ide, IPEI J. 19 (2004) p. 12.
[48] H. Izawa, *Handbook of Fibres (Sen-I Binran)*, The Society of Fiber Science and Technology, Japan, Maruzen Co., 2004, p. 542.
[49] T. Mitsuhashi, *Future Textiles, Protective Clothing from Radioactive Ray*, Sen-I Sha Co., 2006, p. 332.
[50] L. Aubuoy, E-textiles possessing natural thermoregulation properties, Avantex Symposium 2007, AX3.7.
[51] K. Itao, *Technical and Business Development of Electron Cooling Wear*, Extended Abstract of the Lecture of Wearable Computer, Text. Mach. Soc. Jpn. (2006) p. 8.
[52] R. Yokota, *Future Textiles, Space Suit*, Sen-I Sha Co., 2006, p. 313.
[53] Dow Corning, Active protection system: A "smart" impact protection textile, presented at Techtextil Messe 2007 as one of the awarded technologies, 2007.
[54] Pamphlet of 'Quickform', Toyobo Co. Ltd., 2003.
[55] S. Murayama, *Handbook of Fibres (Sen-I Binran)*, The Society of Fiber Science and Technology, Japan, Maruzen Co., 2004, p. 705.
[56] H. Sano, *Handbook of Fibres (Sen-I Binran)*, The Society of Fiber Science and Technology, Japan, Maruzen Co., 2004, p. 712.
[57] K. Kadoya, *Handbook of Fibres (Sen-I Binran)*, The Society of Fiber Science and Technology, Japan, Maruzen Co., 2004, p. 721.
[58] F. Suzuki, *Handbook of Fibres (Sen-I Binran)*, The Society of Fiber Science and Technology, Japan, Maruzen Co., 2004, p. 722.
[59] T. Kaino, *Future Textiles, Optical Computer*, Sen-I Sha Co., 2006, p. 227.
[60] H. Araie et al., Sen'i Gakkaishi 60 (2007) p. 288.
[61] S. Takagi, J. Text. Mach. Soc. Jpn. 58 (2005) p. 434.
[62] F. Baba, *Handbook of Fibres (Sen-I Binran)*, The Society of Fiber Science and Technology, Japan, Maruzen Co., 2004 p. 729.
[63] Pamphlet of 'Toonex', Teijin Fiber Co. Ltd., 2007.
[64] K. Inada, *Handbook of Fibres (Sen-I Binran)*, The Society of Fiber Science and Technology, Japan, Maruzen Co., 2004, p. 728.
[65] S. Matsukawa, *Handbook of Fibres (Sen-I Binran)*, The Society of Fiber Science and Technology, Japan, Maruzen Co., 2004, p. 729.

[66] K. Mizoguchi, *Handbook of Fibrous Materials for Industrial Uses*, The Society of Fiber Science and Technology, Japan, Nikkan Kogyo Co. Ltd., 1994, p. 378.

[67] H. Hoshiro, Sen'i Gakkaishi 63 (2007) p. 94.

[68] V.C. Li, J. Struct. Mech. Earthquake, 2 (1993) p. 37.

[69] M. Ise, *Future Textiles, Concrete Reinforcement*, Sen-I Sha Co., 2006, p. 227.

[70] S. Yanagihara, J. Text. Mach. Soc. Jpn. (2005) p. 437.

[71] S. Onodera, *Future Textiles, Textiles for Reinforcing Earth*, Sen-I Sha Co., 2006, p. 227.

[72] *Profile of Maeda Kosen Products*, Maeda Kosen Co. Ltd., 2000.

[73] A. Hirotsu et al., J. Text. Mach. Soc. Jpn. 59 (2006) p. 620.

[74] T. Akai, *Geosynthetics 1*, Vol. 42, Kakou-gijutsu, Sen-I Sha Co., 2007, p. 105.

[75] T. Akai, *Geosynthetics 5*, Vol. 42, Kakou-gijutsu, Sen-I Sha Co., 2007, p. 473.

[76] N. Yaguchi, *Handbook of Fibres (Sen-I Binran)*, The Society of Fiber Science and Technology, Japan, Maruzen Co., 2004, p. 759.

[77] T. Akai, *Geosynthetics 2*, Vol. 42, Kakou-gijutsu, Sen-I Sha Co., 2007, p. 212.

[78] M. Ishida, J. Text. Mach. Soc. Jpn. 60 (2007) p. 164.

[79] Y. Nakamura, *Handbook of Fibres (Sen-I Binran)*, The Society of Fiber Science and Technology, Japan, Maruzen Co., 2004, p. 771.

[80] T. Sakobe, J. Text. Mach. Soc. Jpn. 60 (2007) p. 489.

[81] J. Maruyama et al., *Handbook of Fibres (Sen-I Binran)*, The Society of Fiber Science and Technology, Japan, Maruzen Co., 2004, p. 763.

[82] Y. Iwasa, *Breathable and Water Proof Sheet Used for Underlining Roof Tile*, Vol. 41, Kakou-gijutsu, Sen-I Sha Co., 2006, p. 157.

[83] H. Minami, *Future of Membrane Structures*, Nikkan Kogyo Shinbun Co. Ltd., 2003, p. 12.

[84] *Paltem Hose Lining*, Home Page of Ashimori Kogyo Co. Ltd., 2007.

[85] G. Franze et al., *Textile Reinforced Multi-layer Tube for Pressure Pipes Systems*, presented at Techtextil Symposium, 2005.

[86] *Electrical Conductive Fibers*, Pamphlet, Toray Co. Ltd., presented at Techtextil 2007 Exhibition.

[87] *Kuralon EC*, Pamphlet, Kuraray Co. Ltd., presented at Techtextil 2007 Exhibition.

[88] H. Nishide, The Extended Abstract of the 6th Nanofiber Technology Symposium for the 21st Century, *Nanofiber world in the organic radical battery*, 2006, p. 1.

[89] Nikkei Bus. (August 29, 2005) p. 134.

[90] Pamphlet of Eleksen Co. at Techtextil/Avantex Exhibition, 2005.

[91] A. Shimizu, *Computer Clothing 6*, Vol. 41, Kakou-gijutsu, Sen-I Sha Co., 2006, p. 411.

[92] Y. Sankai, *Robot Suit 'HAL'*, lecture at Kyoto Institute of Technology, May 31, 2008.

[93] T. Ghosh, Indian J. Fiber Text. Res. 31 (2006) p. 170.

[94] A. Shimizu, *Computer Clothing 5*, Vol. 41, Kakou-gijutsu, Sen-I Sha Co., 2006, p. 345.

[95] K. Igarashi et al., *Positioning system by wearable antenna*, Proceedings of the Annual Conference of Communication Society, The Institute of Electronics, Information and Communication Engineers, 2004, B-3-15.

[96] F. Axisa, *Smart clothes for the monitoring in real time and conditions of physiological and sensorial reactions of human*, Proceedings of the 25th Annual Conference of IEE EMBS, 2003.

[97] A. Ueno et al., *Low invasive measurement of electro-cardiogram for newborns and infants*, Research Report 2003, Frontier R&D Center of Tokyo Denki University, 2003.

[98] J. Hasegawa, *Development of wearable electrode for electro-cardiogram*, Extended Abstract, Lectures of Wearable Computer held by the Textile Machinery Society of Japan, 2006, p. 20.

[99] X. Tao, *Polymeric photonic fibers, smart fabrics and interactive apparel*, Proceedings of the 35th Textile Research Symposium, 2006, p. 51.

[100] W. Chen et al., J. Three Dimens Images 17 (2003) p. 104.

[101] J. Muhlsteff et al., *Wearable approach for continuous ECG-and activity patient monitoring*, Proceedings of the 26th Annual International Conference of the IEEE EMBS, 2004, p. 2184.

[102] P. Gibbs et al., *Wearable conductive fiber sensor arrays for measuring multi-axis joint motion*, Proceedings of the 26th Annual International Conference of the IEEE EMBS, 2004, p. 4755.

[103] R. Paradiso, *Wearable health care system*, Avantex Symposium 2005, Messe Frankfurt, 2005.

[104] H. Kobayashi, Gazo Labo 7 (2005) p. 62.

[105] Nikkei Bus. (November 6, 2006) p. 97.

[106] The news of Yomiuri Shinbun, October 6, 2006.

[107] S. Averett, Ind. Eng. (July 2003) p. 35.
[108] Y. Yamasaki, *R & D of new functional smart working wear using wearable computer*, Extended Abstract, Lectures of Wearable Computer held by the Textile Machinery Society of Japan, 2006, p. 20.
[109] A. Shimizu, *A study on intelligent overalls with wearable computer*, Extended Abstract at Annual Conference of the Institute of Electronics, Information and Communication Engineer, Japan, 2003, A-14-12.
[110] W.B. Zeper et al., *Lumative light emitting textiles*, Avantex Symposium 2007, Messe Frankfurt, 2007.
[111] M. Oe, J. Text. Mach. Soc. Jpn. 59 (2006) p. 552.
[112] Future Mater. 6 (2005) p. 10.
[113] R. Armitage et al., *Commercialization of wearable health monitoring systems*, presented at Avantex 2007 Symposium, 2007.
[114] Pamphlet of Deister Co., presented at Avantex 2007 Exhibition, 2007.
[115] Future Mater. (6) (2004) p. 9.

Bibliography published by Woodhead

General

A.R. Horrocks et al. (eds.), Handbook of Technical Textiles, 2000.
S. Adanur, Wellington Sears Handbook of Industrial Textiles, 1995.

Applications

B.L. Deopura et al. (eds.), *Polyesters and Polyamides*, 2008.
R.A. Scott (eds.), *Textiles for Protection*, 2005.
R.W. Sarsby (eds.), *Geosynthetics in Civil Engineering*, 2006.
W. Fung et al., *Textiles in Automotive Engineering*.
S.K. Mukopadhyay et al., *Automotive Textiles*, Text. Prog. 29 (1999).
D.B. Wooton, *Application of Textiles in Rubber*, 2002.
L. Van Langenhove, *Smart Textiles for Medicine and Healthcare*, 2007.
J.F. Kennedy et al. (eds.), *Medical Textiles 2007*, Proceedings of the fourth international conference on healthcare and medical textiles, 2007.
S.C. Anand et al. (eds.), *Medical textiles and Biomaterials for Healthcare*, 2005.
X.M. Tao (eds.), *Wearable Electronics and Photonics*, 2005.
X.M. Tao (eds.), *Smart Fibres, Fabrics and Clothing*, 2001.

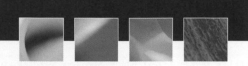

AUTHOR SERVICES

Publish With Us

 Taylor & Francis
Taylor & Francis Group

 Routledge
Taylor & Francis Group

Ψ Psychology Press
Taylor & Francis Group

informa
healthcare

The Taylor & Francis Group Author Services Department aims to enhance your publishing experience as a journal author and optimize the impact of your article in the global research community. Assistance and support is available, from preparing the submission of your article through to setting up citation alerts post-publication on **informa**world™, our online platform offering cross-searchable access to journal, book and database content.

Our Author Services Department can provide advice on how to:

- direct your submission to the correct journal
- prepare your manuscript according to the journal's requirements
- maximize your article's citations
- submit supplementary data for online publication
- submit your article online via Manuscript Central™
- apply for permission to reproduce images
- prepare your illustrations for print
- track the status of your manuscript through the production process
- return your corrections online
- purchase reprints through Rightslink™
- register for article citation alerts
- take advantage of our i*OpenAccess* option
- access your article online
- benefit from rapid online publication via i*First*

See further information at:
www.informaworld.com/authors

or contact:
Author Services Manager, Taylor & Francis, 4 Park Square, Milton Park, Abingdon, Oxon OX14 4RN, UK, email: authorqueries@tandf.co.uk

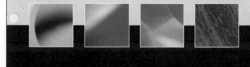

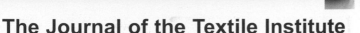

The Journal of the Textile Institute

Published on behalf of the Textile Institute

EDITOR-IN-CHIEF:

D. Buchanan, *North Carolina State University, USA*

The Journal of the Textile Institute welcomes papers concerning research and innovation, reflecting the professional interests of the Textile Institute in science, engineering, economics, management and design related to the textile industry and the use of fibres in consumer and engineering applications. Papers may encompass anything in the range of textile activities, from fibre production through textile processes and machines, to the design, marketing and use of products. Papers may also report fundamental theoretical or experimental investigations, practical or commercial industrial studies and may relate to technical, economic, aesthetic, social or historical aspects of textiles and the textile industry.

SUBSCRIPTION RATES
2008- *Volume* 99 (*6 issues per year*)
Print ISSN 0040-5000
Online ISSN 1754-2340
Institutional rate (print and online): US$599; £315; €479
Institutional rate (online access only): US$569; £299; €455
Personal rate (print only): US$235; £123; €188

A world of specialist information for the academic, professional and business communities. To find out more go to: **www.informaworld.com**

Register your email address at **www.informaworld.com/eupdates** to receive information on books, journals and other news within your areas of interest.

For further information, please contact Customer Services at either of the following:
T&F Informa UK Ltd, Sheepen Place, Colchester, Essex, CO3 3LP, UK
Tel: +44 (0) 20 7017 5544 Fax: 44 (0) 20 7017 5198
Email:tf.enquiries@informa.com
Taylor & Francis Inc, 325 Chestnut Street, Philadelphia, PA 19106, USA
Tel: +1 800 354 1420 (toll-free calls from within the US)
or +1 215 625 8900 (calls from overseas) Fax: +1 215 625 2940
Email:customerservice@taylorandfrancis.com

View an online sample issue at:
www.informaworld.com/JTI